Irrigation
IP
Publishing

P.O. Box 22184
Milwaukie, Oregon 97269-2184

Committed to the advancement of landscape irrigation craftsmanship and technology.

How To Design and Build A Sprinkler System

A Complete Guide For The Do-It-Yourselfer

by

Michael Tenn

Irrigation
IP
Publishing

Milwaukie, Oregon

How To Design and Build A Sprinkler System
A Complete Guide For The Do-It-Yourselfer
by Michael Tenn

Published by:
Irrigation Publishing
P.O. Box 22184
Milwaukie, Oregon 97269-2184

Copyright © 1996, 1997 by Michael Allen Tenn

All rights reserved. No part of this book may be reproduced or transmitted in any form without written permission from the author.

Library of Congress Catalog Card Number: 97-93551
ISBN 0-9657724-3-8

WARNING AND DISCLAIMER

A sprinkler system is a site built project. Because there are so many possible variables from site to site, it is not possible to address every situation that may conceivably arise in the process of building a sprinkler system. This book is intended as a source of general information, and should be supplemented with additional information as needed for your own project. Reasonable effort has been made to make this book as accurate as possible. However, there may be mistakes. The author and Irrigation Publishing will not be responsible for any loss or damage caused or alleged to be caused by the information in this book. Any person who does not accept the above terms and conditions may return this book, unused, to the publisher for a full refund.

ACKNOWLEDGMENTS

Thanks to Shirley Fox, Cheryll Benson, and Vickie Cornell, at Davis and Fox Printing in Gresham, Oregon, for all their help with the book design and production. And, thanks to Brandie Cornell for her help with the drawings.

This book is dedicated to the hard-working men and women in the landscaping industry.
Good people in a fine trade.

❧ TABLE OF CONTENTS ❧

Preface x

Chapter one 1
    The Basics

Chapter two 23
    Design

Chapter three 53
    Low-Volume Irrigation

Chapter four 63
    Construction

Chapter five 85
    Using Your Sprinkler System

Appendix 91
    Pressure Loss Tables and Velocity of Flow Tables

Index 99

Preface

This book is a practical guide for anyone who wants to design and build an intelligent, efficient sprinkler system.

I wrote this book because the thin brochures of instructions, which are sometimes available where retail sprinkler parts are sold, are often incomplete and leave far too many unanswered questions.

At the other end of the continuum are textbooks, written for engineers and irrigation designers. Many of these books are excellent, but they are more technical and expensive than is appropriate for the homeowner project.

For many years I have looked for a book about sprinklers that would truly educate and help consumers. Far too many times, in the irrigation business, I have been called to repair a sprinkler system that was installed by a homeowner and, after seeing it, thought to myself that the only way to repair it would be to rip the whole thing out and start over. I've often wished that there was a book available that would give homeowners the background needed to understand sprinkler systems.

In fact, that brings to mind a question:

What is a good sprinkler system, and how will you know when you have one?

Many of us say that we want our homes and landscapes to reflect "quality", but when it comes to sprinkler systems, do we have any real criteria for evaluating what the characteristics of a quality sprinkler system are? Where can you find the information you would need to answer a question like that?

Maybe I can help.

A quality sprinkler system will have the following attributes:

1. User friendly. Your sprinkler system should not be mysterious and complicated. It should be easy to use.

2. Reliable. You can depend on it, without anxiety, to work as expected when you are sleeping, or at work or on vacation.

3. Constructed with durable, long-lasting components. It should be installed with journeyman level pipefitting skills. The heads, valves, valve boxes. backflow device, piping, controller, and wire should be installed in accordance with accepted industry practices. (This book will explain what those components are, what they do, and how they should be installed).

4. Visually unobtrusive in the landscape. Even though you will do a beautiful plumbing job, when it's all over we want to see plants, not pipes.

5. The components should be readily accessible for service and maintenance. Since you are doing your own work, you will really appreciate the value in this way of thinking.

6. The system will be in compliance with all applicable water quality and plumbing code requirements. A little knowledge up front can save you big headaches later.

7. The sprinklers should provide precise applications of water in a way that is appropriate for the plants, soil conditions and microclimates. It goes without saying that this will conserve water.

8. The system should operate in a flow and pressure range that is consistent with the optimum performance of the system components, and will not exceed the safe capacity of the water service.

9. The sprinklers will apply water with a high degree of distribution uniformity. This is a hallmark of a quality sprinkler system.

10. The system will be designed and built with the highest possible ratio of value to cost. If you are going to spend money on a sprinkler system anyway, why not get the most value possible for your dollars? In a way, that's what this book is all about. The small amount of time and money you invest getting up to speed on sprinkler systems will save you hundreds, and possibly thousands of dollars on your initial installation, and over the life of the sprinkler system.

Sprinkler systems are not so complicated that you couldn't design and build a good one yourself with a little help.

Neither are they so simple that you don't need to do some homework first.

In this book I am going to make available the technical knowledge you absolutely must have to design and build a sprinkler system. And, I hope I will present it to you in a clear, simple, step-by-step way, avoiding any technical information that does not relate directly to homeowner built sprinkler systems.

In addition, I offer you the benefit of my fourteen years of hands-on experience in the field as a licensed landscape contractor who has designed, built, serviced and repaired hundreds of sprinkler systems.

I have taught classes for homeowners, and I have trained professional installers. Those experiences have helped me to develop the material in this volume.

Embarking on a sprinkler project without adequate planning can result in property damage, financial loss, strained relationships and disappointment.

Using the knowledge found in this book, however, I think that you can get an enormous amount of value and satisfaction from the successful completion of your sprinkler system.

A sturdy and well designed sprinkler system can be the source of a great deal of pride, improve the value of your property, save water, eliminate the time consuming and never-ending chore of hand-watering, and make your lawn and garden healthier and more attractive than they have ever been.

I encourage you to read this book in its entirety to get an overview of the design/build process. Then, use the book as a reference as you proceed with your own project.

Good luck with your project. I hope that I can hear from you, or even meet you personally someday, and maybe we can share some sprinkler stories.

Chapter 1

The Basics

In The Basics, *I am going to brief you on the major components of a sprinkler system. An understanding of the essential building materials will be our first step. Without this basic foundation we would not be able to talk about the design and construction process.*

I have endeavored to provide you with as much information as will be necessary for you to be an informed consumer and to get the job done. I have avoided technical details which would detract from the focus of this book, which is intended to be a "Do-it-yourself" guide for the homeowner and not a scientific or technical training manual.

Read through The Basics *first to get some exposure to the terminology of sprinkler hardware. It will make the* Design *and* Construction *sections of the book much easier to understand.*

Don't be overly concerned if you do not completely understand a new concept the very first time that you encounter it. Most of the ideas presented in this first chapter will be expanded on in the rest of the book, and the pieces will fit together as you read on.

Sprinkler Heads

The "end of the line" for water in your sprinkler system piping is a sprinkler head. The sprinkler head is the device that turns the water in the pipe into a spray of controllable shape and size to provide useable water distribution.

In the chapter of this book on design, you will learn how to select and where to place sprinkler heads. In the construction chapter you will learn how to install sprinkler heads. Right now, I will simply explain sprinkler heads in a generic way so that you know what a sprinkler head is and what it does.

Most sprinkler heads consist of a body and a nozzle. There is usually a selection of nozzles that are interchangeable that can be used with any given sprinkler body. The types of sprinklers you will encounter when building your home sprinkler system are as follows:

POP-UP SPRINKLER HEADS

Pop-up Sprinklers are sprinklers that are designed to have the sprinkler body installed in the ground, with the top of the sprinkler head even with the finished grade. When the water is on, a hollow stem (called a *riser* or *flow tube*) rises (pops-up) from the sprinkler body and the water is discharged from a nozzle. When the water is off, the stem and nozzle retract back down into the sprinkler body, out of sight and safe from mower blades and pedestrians. Old style pop-up sprinklers had heavy brass stems and nozzles, combined with loose tolerances between the body and stem, and relied on gravity for retraction. Todays lightweight plastic sprinklers have a strong spring inside the body for a positive retraction of the plastic nozzles. The pop-ups are available in several configurations: Spray, gear driven rotor, and impact. What follows is a brief description of each of the three types of pop-up sprinkler head.

1. Spray head. *Spray heads* discharge water in a fixed, fan-like spray.

The spray head accommodates a very wide range of nozzles. By changing nozzles, a spray head can throw a pattern of water that is shaped like virtually any fraction of a circle including a full circle. For example, if I were to place a spray head in the corner of a square or rectangular lawn, I could select a nozzle that would give me the shape of a quarter circle (90 degrees) to accommodate that head placement and water only the intended area without excessive overspray onto adjacent surfaces, like a driveway or sidewalk.

The same spray head, placed in the center of the lawn, could be fitted with a full-circle nozzle.

There are also nozzles available to accommodate narrow strips where a circular spray would be inappropriate.

Nozzle selection will vary depending on the manufacturer. When you buy sprinklers for your project, compare the nozzle selection that is available for the different brands of sprinkler bodies.

In addition to the shape of the spray, nozzles are selected for the radius, or distance, that they spray. A quarter circle nozzle may be available in versions that deliver distances of anywhere from eight feet to fifteen feet, and these distances may be fine tuned even more if the nozzle has an adjusting screw, which most do.

The other variable in spray heads is the height that the flow tube pops up. In a lawn area, a three or four inch pop-up height is sufficient to direct the spray over the top of the grass. In shrub beds and ground cover areas it is often desirable to have a sprinkler that pops up higher. The greater pop-up height will allow the spray to go over the top of foliage that would otherwise block the spray. Sprinklers are manufactured in a variety of pop-up heights, ranging from two to twelve inches.

2. Gear driven rotor: The *gear driven rotor*, often simply called a rotor, is a sprinkler that has a rotating flow tube and nozzle. The water from this type of head is discharged as a stream which rotates back and forth. Most rotors emit a single stream of water, although some heads have been manufactured which emit multiple streams.

1-1. A pop-up spray head. The flow tube and nozzle pop-up to spray over the foliage.

1-2. When the water is off, a spring inside the sprinkler body (buried below grade) retracts the flow tube and nozzle.

Rotors are powered by water driven gears inside the sprinkler body. The water passing through the sprinkler body is used to power the flow tube rotation as it (water) travels to the nozzle.

Rotors are capable of covering much greater distances then are spray heads. Spray heads are typically spaced not more than 15' apart. Rotor heads in a residential application are often spaced in the 20' to 40' range. Because the rotor is such a popular and versatile sprinkler head, some manufactures have recently introduced rotors designed for smaller areas.

Like spray heads, rotors usually have a large selection of nozzles which are interchangeable with a given sprinkler body. Rotor nozzles are selected according to the amount of water to be discharged and the radius, or distance, to be covered. The Pattern (half circle, quarter circle and so forth) of a rotor is selected by adjusting the *arc*, or amount of rotation, of each individual head.

In addition, most rotors have an adjusting screw in the nozzle to fine tune the distance of throw.

3. Impact sprinkler. *Impact* sprinklers are an older design then rotors. These are the noisy sprinklers that flap back and forth.

Impact sprinklers are powered by the impact of the stream of water that is discharged from the nozzle. The water hits a driver arm that is deflected by the water, and the force of the swinging arm turns the sprinkler. A weight or spring returns the driver back into the stream of water and the process repeats.

1-3. A pop-up impact head.

Impacts have interchangeable nozzles to regulate the distance of throw and amount of water discharged.

The spray pattern is controlled by adjusting the degrees of arc, as with a rotor head.

RISER

A *riser* is used to install any sprinkler that does not pop-up. Because it does not pop-up, the nozzle must be placed on a nipple tall enough to achieve the correct clearance over whatever plants are being irrigated.

A *nipple* is a length of pipe that is threaded at each end. Nipples you use for your home sprinklers could be made of galvanized steel or copper, but will most likely be made from 1/2" diameter, schedule 80 PVC plastic. When a nipple is used for the purpose of mounting a sprinkler, it is called a *riser*.

Risers can be any length you desire, and can be lengthened in sections by connecting nipples together by the use of *threaded couplers*.

Because it is going to be installed on a nipple, above grade, risers are not going to be useable in lawn areas where they could be damaged by mowers, or in areas where they would be a pedestrian hazard, subject to vandalism, or be a visual blight.

Just about any spray head nozzle can be mounted on a riser by using a *shrub adapter*. A shrub adapter is a fitting that works as a transition from the threads on the nozzle to the threads on the riser.

You could mount the entire pop-up sprinkler body on a riser. But, if you are using a riser there is no reason to pay for a sprinkler body. A shrub adapter will be less expensive and much lighter in weight then using the entire sprinkler body.

Impact sprinkler nozzle assemblies can also be mounted on risers without using the bulky pop-up body.

Rotor nozzles cannot be mounted on a riser without the rotor body. That is because the body of a rotor contains the gears necessary to make the nozzle turn. You can mount a rotor on a nipple, but you need to mount the entire body and nozzle assembly.

Sprinkler head selection and installation details are provided in the chapters on design and installation.

Valves

Valves are the component in the sprinkler system that start and stop the flow of water. There are different types of valves used in a sprinkler system.

In a manually operated sprinkler system, all of the valves are operated by hand.

In an automatic sprinkler system, the valves which regulate the flow of water to the sprinkler heads open and close electrically.

Valves are made from either plastic or brass.

Valves have several different functions in an irrigation environment.

THE FUNCTION OF VALVES IN A SPRINKLER SYSTEM

Main valve: The *main valve* is the valve installed at the point of connection between your water supply and the sprinkler system. When the main valve is open, the sprinkler mainline is charged (filled with water under pressure). Usually, the main valve remains

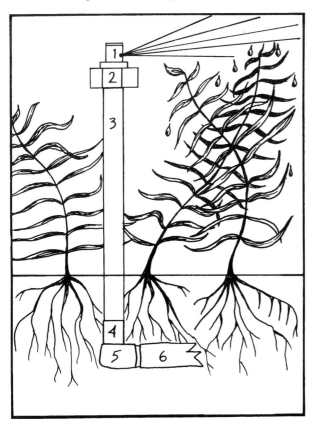

1-4. A spray nozzle mounted on a nipple (riser). KEY: 1. Spray nozzle. 2. Shrub adapter. 3. 1/2" diameter, schedule 80 PVC nipple. 4. 1/2" diameter schedule 40 PVC threaded coupler. 5. 1/2" thread x barbed 90 degree ell. 6. 1/2" diameter PE pipe- connects to PVC lateral line.

fully open throughout the operating season of the sprinkler system. The main valve allows you to shut-off the flow of water to the entire sprinkler system without interrupting your household water service. The mainline is the piping between the main valve and the control valve(s). When the sprinklers are not operating, the mainline remains charged (pressurized) at the same static pressure as the other household piping, as long as the main valve remains open.

Control valve: A *control valve* is a valve that controls the flow of water between the mainline and the *lateral* lines which feed the individual sprinkler heads. *Lateral* piping is the name given to the piping on the discharge side of the control valves.

In a typical sprinkler system, the main valve is opened in the spring and remains open throughout the watering season, so the mainline is under constant pressure. The control valves, which are attached to the mainline, only open when it is time to water. Therefore, the lateral piping is only pressurized when the sprinklers are actually running.

Drain valve: A *drain valve* is a valve which can be opened to drain water from the sprinkler piping. This is useful in a climate where leaving water in the pipes could result in freeze damage during the winter months.

Valves in a residential sprinkler system function as either main, drain, or control valves. You have some latitude as to the type of valve you choose to perform those functions.

Types of Valves Available for Home Irrigation

Manual valves: *Manual valves* are non-electric valves which are operated either by hand or with a hand held sprinkler key. The manual valves you will consider are:

Gate valve: When the handle of a *gate valve* is turned counter-clockwise, an internal "gate" is lifted and water flows through the valve. Gate valves are frequently used as a main shut-off valve because when fully open they produce little restriction to the full flow of water. Gate valves are most commonly available with a "wheel" handle that is easier to turn by hand than with a sprinkler key. Gate valves seem to lose their ability to remain 100 percent drip tight over time because of wear between the gate and the "seat" where the gate rests when the valve is closed. Choose a gate valve with a "resilient seat" for long life.

Gate and waste valve: A *gate and waste* valve is a gate valve with the addition of a drain plug installed on the "downstream" side of the gate. When the gate is closed, the plug can be loosened to drain the water from the piping on that side of the valve. This can be a desirable feature in a freezing climate, where sprinklers are winterized by removing residual water from the piping.

1-5. A gate and waste valve. The knurled cap on the side of the valve can be loosened to drain water from the piping on the discharge side of the valve when the valve is shut off.

Globe angle valve: A *globe angle valve*, frequently called simply an angle valve, is a valve that uses a stopper instead of a gate to control the flow of water. Water entering the bottom of the valve is held back by a stopper with a rubber disk, which seals against a machined valve seat. When the valve handle is turned counter-clockwise, a valve stem lifts the stopper (or disk holder as it is called) and the rubber disk off the valve seat, allowing water to flow through the opening and out the discharge side of the valve.

Angle valves are most frequently used as the control valves in manually operated sprinkler systems, and usually come with a "cross" handle which accommodates a sprinkler key. This feature lends itself to a valve that is usually installed below ground and requires a long-handled key for operation.

The design of an angle valve is such that it also lends simplicity to installation where the mainline piping is installed at a greater depth than is the lateral piping, as is usually the case. A vertical nipple connects the angle valve to the mainline, and the angle valve discharges the water to a horizontal lateral line.

When an angle valve is installed above ground, it is frequently in a configuration that is manufactured as a combination of angle valve with an atmospheric vacuum breaker, also called an anti-siphon valve. This will be covered in the section on backflow prevention.

An angle valve may also be installed to function as a drain valve.

One variety of angle valve is manufactured so that the discharge side of the valve is configured to accommodate the attachment of a garden hose. This valve is usually installed above ground on a nipple, and is referred to as a *garden valve,* although you may also know it as a faucet or hose bib.

1-6. Two configurations of the globe angle valve. The valve on the left has a female pipe thread on both the inlet (bottom) and discharge sides. The valve on the right is also an angle valve, but it is configured to accept a hose thread on the discharge side, making this a "garden valve".

Globe straight valve: A *globe straight valve,* also referred to as simply a *straight valve,* uses the same stopper mechanism as the angle valve to control the flow of water. the difference is that the intake and discharge openings of the straight valve are on the same horizontal plane, just like the gate valve. Straight valves are most often used as manual drain valves.

The rubber disk in the globe type valves is replaceable when it wears out by removing the top *(bonnet)* of the valve. For this reason, globe valves are often preferred to gate valves in applications where frequent opening and closing of the valve is required.

Ball valve: A *ball valve* consists of a valve body with a stopper that looks like a ball with a large hole drilled through the center. When the handle of a ball valve is turned 90 degrees in either direction, the ball inside the valve body rotates, allowing the water to pass through the hollow center of the ball. When the handle of the ball valve is rotated another 90 degrees in either direction, the solid side of the ball blocks the flow of water.

Ball valves are a high quality shut-off valve, and are most frequently seen as the shut-off valves for backflow devices. A ball valve makes a good main shut-off valve except that the handle of a ball valve does not lend itself to being easily turned with a valve key if the valve is installed underground.

Quick-coupling valve: A *quick-coupling valve* is a valve installed on the mainline that remains closed until activated with a *quick-coupling key.* A quick-coupling key is a hollow lug inserted in the keyway of the quick-coupling valve. In residential applications, quick-coupling valves are most often used as a place to connect an air hose in climates where compressed air is used to purge water from the sprinkler pipes in the winter.

Automatic drain valve: An *automatic drain valve* is a valve that relies on the water pressure in the piping to stay closed. When the pressure in the pipe drops below a designated pressure, the valve opens and water in the pipe drains out. If an automatic drain is installed on a lateral line, the valve will close when the sprinklers are running and open when the control valve shuts the sprinklers off. That will drain the lateral line every time the sprinklers shut off. Automatic drain valves are most useful in climates where the temperature can

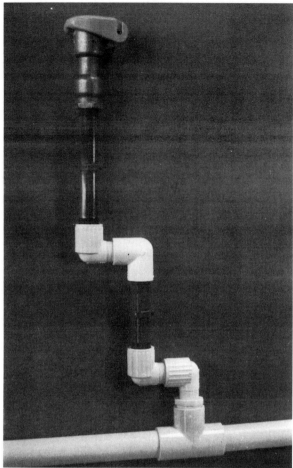

1-7. A quick-coupler valve assembled on double swing joints

fluctuate from warm to below freezing and back to warm in a brief time. Colorado comes to mind as a climate where you may need the sprinklers on because of warm weather, and the next day get hammered with a snow storm before you can winterize the sprinklers. In a climate like that, automatic drains can be a useful feature.

ELECTRIC VALVES

Electric valves are valves which open and close by a mechanism that is activated by applying voltage to a solenoid. It is the electric valve that allows us to use a controller to turn the sprinklers on and off automatically.

The typical residential electric valve consists of a rubber *diaphragm* inside a plastic valve body. The diaphragm is designed so that the top of the diaphragm has more surface area then does the underside of the diaphragm.

The valve is ported so that water entering the valve flows into a chamber above the diaphragm and also beneath the diaphragm. Because of the greater surface area above the diaphragm, water pressure holds it in the closed position, tight against a seating surface. There is also a spring between the top of the diaphragm and the underside of the *valve bonnet*.

The *solenoid* on the top of an electric valve contains a small plunger which seals a very small hole called an *exhaust port.* When the controller sends a 24 volt current to the solenoid, it causes this plunger to lift and water from the top of the diaphragm escapes through the exhaust port. The loss of pressure that results allows the water on the bottom of the diaphragm to overcome the spring tension, and the diaphragm lifts off the valve seat, causing water to flow through the valve to the downstream piping. The valve is then open.

When the controller stops sending current to the solenoid, another small spring returns the plunger to its position of sealing the exhaust port. With the exhaust port closed, the top of the diaphragm again builds a pressure exceeding the pressure on the underside of the diaphragm , and with the additional force of the spring, pushes the diaphragm back down to the valve seat, closing the valve.

When an electric valve has a *latching solenoid* it means that the solenoid can remain open without continuous electrical input. This is an energy saving feature.

It is possible to cause an electric valve to function manually by finding out how to bleed water off of the top of the diaphragm. this is usually accomplished with a bleed screw mounted somewhere on the bonnet, or top of the valve , or by turning the solenoid itself. Some valves have a small lever under the solenoid that is a manual bleed. The manual bleed feature is useful when you want to activate an electric valve without going to the controller.

Some electric valves also have a *flow-control* feature which allows you to regulate the flow of water through the valve. Without

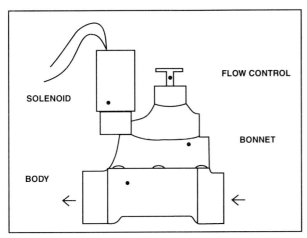

1-8. A typical electric control valve. The two wires on top of the solenoid will be spliced into the "field wires."

this feature, you will have a valve that is either completely open or completely closed.

Electric valves are typically installed underground and are located in plastic valve boxes which provide access to the valves for servicing.

When an electric valve is installed above ground, it is usually in a configuration that is manufactured as a combination electric control valve and atmospheric vacuum breaker (anti-siphon valve).

Valve selection and installation will be explained in the sections on design and construction.

❧ Valve boxes ❧

Valve boxes (sometimes called valve pits) are plastic containers that are installed to contain the underground valves. A manual valve installed underground requires access from the top so that it can be reached by a valve key or by hand. An electric valve installed underground requires access from the top for service and adjustment.

Anyplace where a group of wires is spliced, it is also advisable to bundle the waterproof splices together in a valve box. In the event of any future electrical troubleshooting, knowing where the splices are located can be a great help.

Valve boxes with lids are manufactured in several standard sizes. Round boxes are available in 6" and 10" diameters. Rectangular boxes are available in approximately 18"x12"x12" high and 24"x18"x12" high. Of course, the exact dimensions will vary slightly on products from different manufactures.

In addition to valve boxes, there are several manufactured products available as caps to cover the end of 2" and 4" diameter pipe so that the pipe can be cut to length and used as a sleeve reaching down to underground valves. Three-quarter inch and 1" manual valves with a cross handle can be sleeved with 2" PVC pipe. Most manual valves up to 2" with a wheel handle can fit into a sleeve made from 4" PVC pipe.

A 10" round valve box will accommodate a single 1" electric valve and a 18" x 12" x 12" rectangular box will hold up to three 1" electric valves or a single 1-1/2" or 2" electric valve.

A valve box for electric valves should be sized so that the top of the valve(s) and solenoid(s) can be removed for service without digging out the box. Putting valves in a box that is too small will make any future repair difficult.

A sleeve for a manual valve needs to be large enough to turn the handle of the valve without binding against the sleeve.

Any valve, manual or electric, may need access at some time. For this reason, you should always provide a box or sleeve for every valve, and then be sure that the lid or cap remains visible. Do not allow the top of a valve box to be covered by mulch, a berm, groundcover, a deck or anything else.

Occasionally I will see a situation where a homeowner has added a patio, walk or driveway after the sprinklers were installed, and has surrounded a valve box or sleeve with concrete. Just suppose you have a manual valve with a cross handle in a 2" sleeve surrounded by concrete and the packing nut on the valve starts leaking. All you need to do to repair the valve is to tighten the packing nut a little bit. Unfortunately, in this situation, you'll need to remove the concrete with a jackhammer or sledge before you can dig out the valve enough to get a wrench on it. After the valve is fixed you will need to repair your concrete. What should be a simple adjustment becomes an expensive ordeal. For that reason, don't ever surround valves with a permanent hardscape. If you are remodeling a yard with an existing sprinkler system, you will be better off relocating any valves that are in the way of a proposed hardscape.

Valve box installation is detailed in the construction part of this book.

Pipe

I am going to discuss the types of pipe that you are most likely to encounter when building a home sprinkler system.

Pipe may be manufactured from a variety of materials. The most common types encountered in sprinkler systems are plastic, copper and galvanized steel. In this section of the book you will learn about the general descriptions and uses of the different types of pipe, as well as some basic pipefitting instruction.

Plastic

Polyvinyl Chloride (PVC)

Polyvinyl Chloride (PVC) is the most common plastic pipe in use today in the construction of home irrigation systems. PVC is strong, lightweight, durable and easy to work with. The walls of PVC pipe are very smooth and provide the ability to move water with a minimum of pressure loss from friction.

PVC pipe comes in various diameters. Selecting the correct diameter of pipe for a given volume of water is a design consideration and is addressed elsewhere this book. PVC pipe commonly comes in 20' lengths.

PVC pipe is rated for strength. For example, class 160 is lighter (thinner walled) than class 200 and has a lower bursting strength. Lighter pipe is less expensive than the heavier pipe. For most sprinkler systems, class 200 pipe is adequate for piping on the discharge side of the backflow device. Piping on the upstream side of the backflow device may be mandated by a local plumbing code to be of a heavier material, such as schedule 40 PVC. Check with local suppliers to find out what is required and customary in your area. Anytime PVC pipe is used under a sidewalk or driveway it should be schedule 40.

The plastic fittings (couplers, tees, ells and so forth) that you will use to assemble the pipe will be made of schedule 40 PVC.

Schedule 80 PVC is used to manufacture nipples. Nipples are lengths of pipe which are threaded at each end. Schedule 80 is a very strong, heavy-walled plastic. It is gray in color.

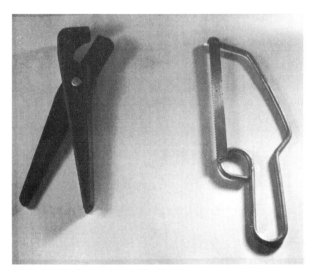

1-9. Tools for cutting plastic pipe.

Cutting

There are a number of pipe cutters on the market for cutting PVC pipe. I like to use a simple hacksaw with a 6" blade. The hacksaw works fine and will cut pipe that is already in a trench in positions where extra digging would be needed to get all the way around the pipe with the handle of a pipe cutter. The hacksaw also works better when the pipe to be cut is laying next to another pipe- again because you can cut from one side without needing to get a handle all the way around the pipe.

Make your cuts square to the length of the pipe. A pipe that is cut at an angle won't bottom out completely when it is inserted into

1-10. Cutting PVC pipe with a 6" hacksaw.

the fitting, and so won't form as strong a joint as a pipe that is cut square.

Cleaning

The cut ends of the PVC pipe will have burrs. There are several deburring tools available. You could also use a pocket knife to lightly clean off the burrs. The tool that works best for me is my thumb and forefinger. Aside from getting the burrs off, wipe any dirt and moisture off the pipe with a rag and it's ready to glue.

Joining

To join plastic pipe, swab the outside of the pipe and the inside of the fitting with primer and then put PVC glue on both parts. Apply the glue evenly with the dauber, going completely around the outside of the pipe and the inside of the fitting. Quickly insert the pipe in the fitting and give it a twist to spread the glue evenly. Hold the pipe and fitting together firmly for a few seconds.

Putting a bead of glue around the outside of the fitting where it meets the pipe does no good at all. In fact, I use a rag to wipe off the excess glue that squeezes out, because that glue will soften and weaken the outside walls of the pipe.

There are different types of glue available. I like to use a medium bodied, fast drying, blue glue that contains primer. This glue can be used without the use of a separate primer.

Because I use this type of glue, I only use a separate primer when I'm working with pipe or fittings that are very dirty; when I'm using pipe that is over 2" in diameter; and on the upstream side of a backflow device where the plumbing code will require a primer.

Drying times for the different glues will vary. Follow the instructions on the can.

POLYETHYLENE (PE)

Polyethylene pipe (PE) is a soft-walled, flexible pipe that is sold in rolls. This pipe is often used in the construction of sprinkler systems in colder regions. PE pipe is not as smooth-walled as PVC pipe, and so produces more pressure loss from friction when the water is flowing. PE is also more subject to aging and then is PVC. PE pipe will have a lower pressure rating than PVC pipe. Being flexible, however, means that the PE pipe will be less likely to crack from freezing than will PVC.

In some areas, it is the custom to use PVC pipe for mainlines in conjunction with PE pipe for lateral lines.

A very popular product is a 1/2" diameter PE flexible tubing that is widely used to connect sprinkler heads to PVC lateral lines.

The tubing used in low-volume irrigation systems is also made from PE pipe.

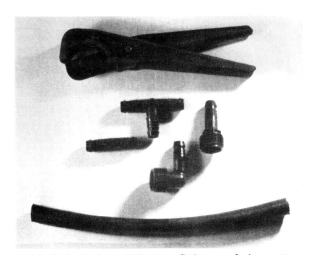

1-11. Polyethylene (PE) pipe, fittings and pipe cutter. This flexible 1/2" pipe is frequently used to connect sprinkler heads to PVC pipe. Larger diameter versions of this pipe are used instead of PVC pipe in some regions.

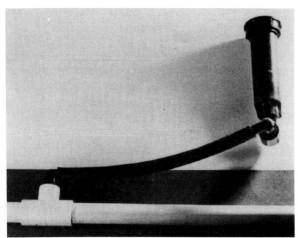

1-12. Using PE pipe to connect a sprinkler head to a PVC sprinkler line.

Cutting

PE pipe is easily cut to length with a pipe cutter, hand pruners, a knife, or a hacksaw.

Joining

The fittings for PE pipe have barbs, or ridges, to hold the pipe onto the fitting. The barbed end of a fitting is inserted in the pipe, and then a clamp is tightened over the outside

1-13. Cutting PE pipe.

of the pipe and fitting to hold it in place. There is a crimping tool available that will facilitate putting the clamps on.

The popular 1/2" PE flexible tubing which is often used to attach sprinkler heads to a lateral line uses barbed fittings which require no clamp. It is important, however, to make sure that the flexible tubing is pushed completely over all the ridges on the fitting to insure a secure fit.

COPPER PIPE

Although you will not be building your sprinkler system from copper tube, you may find that the water line supplying your home from the water meter is made from copper. If so, then you will be working with copper tube to make your point-of-connection between that water line and the main shut-off valve that you will install to feed the sprinkler piping.

You may also encounter a need to work with copper tube if you want to route any of your piping through a basement or garage where a building code might prohibit plastic pipe.

Copper tube comes in different weights. The wall thickness of copper tube is specified as being either type "M", "K" or "L". Check to see what type is recommended in your area for your specific application.

Cutting copper tube

Copper tube will be cut with a tubing cutter which consists of a cutting wheel that is moved 360 degrees around the tube until it is cut completely through. When placing your tube cutter on the copper tube, do not squeeze down so tightly that you dent the copper tube. Copper tube that is pinched, or out-of-round, will not produce a leak free joint.

Copper tube may also be cut with a hacksaw when there is no room to use a tube cutter. Cutting with a hacksaw will leave sharp burrs on the tube which should be carefully removed with a file.

As with all pipe, the cut should be made at a right angle to the length of the tube.

Joining copper tube:

1. Clean all the surfaces to be joined. The outside of the tube and the inside of the fitting(s) should be cleaned with emery cloth until they shine. Having the tube and fittings absolutely clean and dry is critical.

2. Apply a paste flux to the outside of the tube and the inside of the fitting(s) and fit the pieces of your assembly together. Figure out how you are going to keep the pieces from moving while you are sweating the joints

1-14. Tools and supplies for sweating copper pipe: Emery cloth, regulator, propane, tube cutters, flux, striker and lead-free solder, approved for use on potable waterlines.

together, given the fact that the copper will be much too hot to hold in place with a hand, even with gloves.

3. Apply heat evenly with a propane torch, moving the flame back and forth on the fitting as you move it 360 degrees around the outside of the tube and fitting.

4. As the flux burns off and the fitting reaches the right temperature, touch the end of a length of lead-free plumbing solder to the crack where the tube and fitting meet. If the heat is right, the solder will easily be sucked into the fitting by capillary action, even if it is traveling upward. If the solder does not easily take to the fitting, pull the solder away and continue heating the tube and fitting and then try again. Do not try to make the solder melt by putting the flame directly on the solder. Instead, let the heat of the fitting pull the solder in. When the fitting is taking up the solder, move the solder all the way around the fitting while directing the flame at the other end of the fitting.

Once the crack has been filled with solder all the way around, remove the heat and move on to the next joint. Excess solder will run back out and can interfere with other joints or run onto threaded fittings.

Sweating copper tube is one of those things that requires a little practice. My suggestion is that if this is new to you, put some copper tube in the vice on your workbench and sweat a few fittings until you are comfortable with the process before you try the real thing.

When difficulty is encountered making a sweat joint work, it is usually because either the tube and fitting have not been cleaned adequately, the tube is out-of-round, or there is water in the tube.

When you are sweating on a water line that may contain some water, remember that you can create steam inside the tube that will build up pressure and burst if an opening is not provided for the steam to escape.

> If you have a problem with dripping water in the tube ruining your efforts, here's a *trick of the trade:* You can hold back a small amount of dripping water temporarily by balling up some white bread and stuffing it in the pipe as a dam. This will hold the drip back long enough to sweat in a tee. When the water is turned back on, the white bread will dissolve and can flush through the piping with no ill effects.

GALVANIZED STEEL PIPE

You will not be using steel pipe in your sprinkler system except possibly for steel nipples on which to install a pressure vacuum breaker or an atmospheric vacuum breaker. The other place where you may encounter steel pipe is when you do the tie-in to your water line. Some homes, especially older homes, will have a galvanized steel water line.

All steel pipe becomes corroded with time. When you uncover a steel water line, carefully knock away the outer scale with a hammer and chisel or screwdriver and brush it with a wire brush. Sometimes steel pipe has a lot of rust built up on the outside, but when the pipe is cleaned it is sound underneath. Then again, the pipe may be corroded badly and the walls of the pipe could crumble under the force of a wrench or a pipe threader. Anytime you encounter old galvanized pipe, there is the chance that the pipe may be too badly corroded to be workable.

Before you cut into a galvanized water line to connect your sprinkler system, understand that if the pipe is so weak and corroded that it cannot be re-threaded, then you will probably need to replace the entire water line from the meter to your house.

If you live in an older home with a galvanized waterline, it is just a matter of time until you need to replace your waterline anyway. If the flow of water from your plumbing fixtures is weak due to corrosion build-up on the inside of your waterline, you may want to consider that this is a good time to get that project done, also. Your yard is going to be trenched up for the sprinklers anyway, why not install a new copper or plastic waterline for the house? The waterline replacement is beyond the scope of this book. If it is something that you want to do, you will need to get a plumbing permit and have it inspected. If you are not comfortable doing it yourself, you can always call a plumber. The main point I want to make here is that a corroded steel waterline may not be workable to install a tee for the sprinkler system, and once you cut into an old and badly corroded pipe you may have no way to get it back together. If that happens then you will have to replace your entire waterline immediately or you will not have water to your house. So, once you have determined that you have a steel waterline, be prepared to replace it quickly if you have to. That might mean getting some bids from plumbers in advance so that if you have to do it you will already know who to call and how much it will cost.

If the galvanized steel pipe is sound, and it often is, you can proceed with your point-of-connection and install the irrigation main (shut-off) valve.

Cutting steel pipe

To cut steel pipe you will want to have an electric reciprocating saw with a steel cutting blade. You can cut steel pipe with a hacksaw, but it is a chore. If you are going to cut steel pipe with a hand saw, use at least a 12" blade for a 1" diameter pipe. The 6" blade that works so well on PVC pipe will be very labor-intensive to cut steel with.

Joining steel pipe:

Steel pipe is joined with threaded fittings. Therefore it is necessary to cut threads in the end of the pipe after it has been cut to size. To accomplish this you can rent a ratchet handle that accepts a variety of cutting die sizes and easily cut the threads as you need to by inserting the appropriate die over the end of the pipe and using the handle to turn the die and cut the threads.

When you are joining steel pipe with threaded fittings, use a pipe dope to help seal against leaks, lubricate the threads, and protect the steel from rusting.

When you need to take apart threaded fittings that are rusty, always work some penetrating oil into the threads before using your pipe wrench.

An aside on pipefitting:

Small diameter plastic pipe is very forgiving. If a pipe is cut a little too short or too long it can be pushed and pulled and bowed to make it fit. Copper pipe cannot be manhandled like plastic can, but it does give a little. Steel pipe won't budge. When you work with steel pipe it is well to remember the adage: "Measure twice and cut once."

❧ FITTINGS ❧

Fittings are the pieces that are used to connect the pipes and valves of the sprinkler system together. The fittings have descriptive names. For example, a *90 degree ell* is an elbow that joins two pipes at a 90 degree angle. A *tee* is a T shaped fitting that joins three pipes.

Fittings are sized by the outside diameter of the pipe they are used with. For example, a *3/4" 90 degree ell* is used to join two lengths of 3/4" pipe at a right angle. The outside diameters of PVC and steel pipe are the same and may be specified with the "iron pipe size" (ips) designation.

Copper pipe has a smaller outside diameter and is specified with the "copper pipe size" (cps) designation. Here is a list of some of the jargon of fittings that will help you to identify and to specify the fittings you need:

- MIPT - Male iron pipe thread. The standard thread. Iron pipe and plastic pipe have the same dimensions. Therefore, the fittings for PVC pipe are sometimes specified as the nominal size plus the designation "ips" for iron pipe size or "ipt" for iron pipe thread.
- FIPT - Female iron pipe thread. The standard thread on a female part.
- CPS- Copper pipe size. The dimensions of copper parts (outside diameter) are one size smaller than the equivalent part in iron or plastic.
- S - Slip. The female socket on a plastic part.
- SPGT- Spigot. A male part sized to be inserted into a female socket.
- C - Compression.

The fittings used in a sprinkler system will join the components by the use of either a slip joint, threads, barbed insert, or compression. When it is necessary to go from one method of joining to another, a fitting is used that has one type of joint at one end and the other type of joint at the other end. For example, a *male adapter* is a fitting that has male threads at one end, and a slip joint socket at the other.

Sometimes you will find it necessary to join unlike materials. For example, you may want to connect plastic pipe to a brass valve. Since you can't glue the plastic to the brass, you will use a plastic *male adapter* which can be threaded into the brass threads of the valve. Then, the plastic pipe can be glued into the socket end of the same male adapter.

Unions are a type of fitting that can be used in places where a particular component of the sprinkler system may need to be removed periodically for service or replacement. The value of a union is that a component installed inline can be replaced without cutting the pipe.

Compression fittings join two pipes together by the use of a coupler with two end caps. One end cap and the coupler slip over the outside of one pipe and the other end cap is slid onto the other pipe. The coupler is then centered over the gap between the pipe ends, and the end caps are tightened down, forming a seal with rubber washers.

Another type of compression fitting is used in low volume irrigation. In that application, flexible PE pipe is inserted into a fitting that is designed so that the pipe cannot be pulled back out.

✿ Sleeves ✿

Whenever pipe and/or wires need to cross under a permanent hardscape surface such as a sidewalk or driveway, it is advisable to provide a *sleeve*. A pipe used for sleeving purposes should be at least two sizes larger in diameter than the pipe it will contain. For example, if I want to install a sleeve under a driveway to accommodate a 3/4" pipe, I will use a minimum 1-1/4" pipe for my sleeve and not a 1" pipe. The reason is because the belled end of a length of 3/4" pipe will not fit through a 1" pipe. One inch pipe needs a sleeve that is at least 1-1/2" and so on.

The best time to install sleeves is before the concrete or other hard surface is installed. If you are building a new home, have your builder or landscape contractor put a sleeve under the driveway and walks before concrete

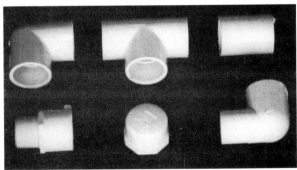

1-15. Schedule 40 PVC fittings. From left, top: Ninety degree ell (slip x slip), tee (slip x slip x slip), coupler (slip x slip). From left, bottom: Male adapter (male thread x slip), Cap (female thread), ninety degree street ell (spigot x slip).

is poured. Make sure the ends of the sleeves have some plastic taped over them to keep the dirt out before they are buried, and mark the location of the sleeve ends with a stake so you can locate them later. You can also make a note of where the sleeves are located relative to some fixed point for a later reference.

If you need to get a pipe under a walk or driveway that is not sleeved, you can bore under it or cut and patch the concrete. My first choice, however, would be a sleeve installed before the hardscape is built.

Controllers

The controller is the "brain" of the automatic sprinkler system. In a manually operated sprinkler system, there is no controller. You turn the valves on and off by hand. In an automatic sprinkler system, the controller, installed in conjunction with electric solenoid operated valves, turns the valves on and off automatically.

The controller will tell the valves what days to water, what time to water and how long to water.

Controllers are programmable. Days, times and duration of the watering cycle can be changed as often as necessary to adjust for the seasonal variation of water requirements.

There are three types of controller:

1. Electro-mechanical controllers. *Electro-mechanical controllers* are constructed of gears and electric timing motors. Programming is done by using knobs, push-pull pins and/or sliders.

Electro-mechanical controllers are simple to use because of their straight- forward mechanical operation.

Programming is relatively limited with this type of controller, but can be completely adequate for a simple watering schedule. Electro-mechanical controllers are not as vulnerable to small fluctuations in the power supply as are solid state controllers.

2. Solid state controllers. *Solid state controllers* use circuit boards, computer chips and digital displays instead of gears and motors. A solid state controller is programmed by using a keypad.

The solid state controller is much more compact than the electro-mechanical controller and more powerful in terms of being able to control more complex watering schedules.

Because a solid state controller stores programming electronically, it is vulnerable to program loss caused by an interruption of the power supply. For this reason, most solid state controllers have a nine volt battery back up to protect the memory.

Solid state controllers give you the ability to use multiple programs. For example, you might have the lawn watering four days a week on one program, the shrubs watering two days a week on another program, and have a drip system watering container plants every day.

Solid state lets us put together a watering schedule that is much more precise for meeting the needs of individual plant groups and microclimates.

The downside to the early solid state controllers was that they were not user-friendly and many homeowners had trouble programming them. This led to the development of the hybrid controller.

3. Hybrid controllers. A *hybrid controller* is a solid state controller that uses knobs, sliders and push buttons to set the program, instead of a keypad. This makes these controllers more user-friendly then previous solid state controllers were. The ability to handle complex watering schedules, combined with easy programming, has made the hybrid design a popular residential controller.

All controller types come in both indoor and outdoor versions.

The outdoor versions come in a weather-proof cabinet.

All controllers have a transformer which converts your 117 volt AC household current to 24 volt AC current to run the low-voltage solenoids in the sprinkler valves. The transformer also puts out 9 to 11 volts to run the memory circuit on solid state controllers.

An indoor model controller will often have an external transformer that is integrated into the wall plug and plugs into a normal household electrical receptacle.

In a outdoor model controller, the transformer will be located inside the weatherproof cabinet, and needs to be hardwired into the 117 volt AC power source.

Controller selection and installation will be explained in the sections on design and construction.

Moisture Sensors

Moisture sensors take automatic watering one step further by interrupting the automatic watering cycle if moisture is already present because of rainfall.

Moisture sensors may be soil probes which are inserted into the ground at strategic locations, or may be of a type that is installed above ground and is triggered by an accumulation of water in a catch device.

Moisture sensors can help to avoid the spectacle of an automatic sprinkler system watering during a rainstorm.

Wire

The electric solenoids on the sprinkler valves are operated by the low-voltage (24 volt AC) power supplied by the transformer in the controller. To get this power from the controller to the valves you will use direct burial wire.

Most residential sprinkler systems will use either 14 gauge single strand insulated copper wire (14-1) or multiple strand 18 gauge wires.

Whenever wire is spliced, the wire nuts should be placed inside moisture resistant containers which are manufactured for this purpose.

In some locations you may be required to obtain a low-voltage electrical permit before doing your wiring.

Wire selection and installation will be explained in the sections on design and construction.

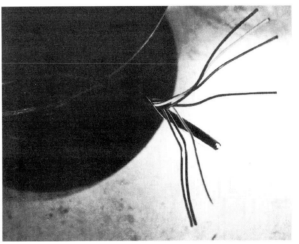

1-16. Low-voltage direct bury wire. This is 18-7. It has seven 18 gauge insulated copper wires bundled in one outer sheath.

Backflow Prevention

Backflow is a condition that can potentially occur whenever a water user connects to a home's water supply for a non-potable water use. When a backflow condition occurs, the normal direction of flow in the affected piping is reversed, making it possible for contaminated water to flow back into the drinking water supply. There are a number of hydraulic anomalies that can, and do, create backflow conditions.

The plumbing fixtures in your home already have backflow preventers that you may not even be aware of. For example, very old bathtubs used to have a fill inlet located below the rim of the tub. This made it possible for the inlet to become submerged when the tub was full. With the inlet submerged, used bath water could, under certain conditions, flow back into the drinking water. If you take a look at the inlets to your sinks and tubs today, you will notice that the inlets are required to be a prescribed distance above the rim of the fixture, so that there is an air gap separation between the flood rim and the inlet. That air gap is, in fact, the required backflow preventer for those fixtures.

Over the years, as sources of the contaminants in our water supply have been identified, steps have been taken to insure the safety of our drinking water. Just as the design of household plumbing fixtures has evolved, so has the design of sprinkler systems. Many older sprinkler systems were installed without any backflow prevention. That does not mean that you should not provide backflow prevention on yours. I don't personally know of any water district that does not require a backflow device on a sprinkler system.

What follows is a brief description of each of the four types of backflow device most commonly used with irrigation sprinkler systems.

1. Atmospheric vacuum breaker (AVB). The *AVB* is potentially the lowest cost form of backflow preventer. It is also the device most often incorrectly installed

The AVB is required to be installed a minimum of 6" above the elevation of the highest head in the zone on which it is installed. The AVB is to be installed on the downstream (discharge) side of **each** control valve in the sprinkler system, and no valves are allowed to be located on the downstream side of an AVB.

The body of an AVB contains a poppet and a vent. When the water is flowing, the poppet rises and seals the vent so that water flows through the body of the device to the outlet. If the pressure on the upstream, or supply side, of the AVB drops, then the poppet falls and the vent opens. Atmospheric pressure then breaks the siphon which would otherwise be created by a negative pressure, and stops the reverse flow of water in the piping. Additionally, any water which does come back will now spill out of the open vent as the poppet seals the inlet side of the device.

In this way, the AVB can protect against a reverse flow of water caused by siphonage, but the device does not protect against a reverse flow caused by backpressure.

An AVB is not meant to be operated under constant pressure, therefore it is not suitable for installation on a mainline. One of the most common mistakes homeowners make is to install an AVB on a mainline. The other backflow devices we will discuss are appropriate for mainline installation. The AVB, however, is only intended to be used in conjunction with control valves and on lateral piping.

A popular homeowner valve is the combination control/anti-siphon valve which incorporates in its design a control valve and an AVB in one integrated unit.

Cost-wise the AVB can be the least expensive device to install, per device. But, since you will need one device for each zone, the savings can be negligible on a system with more than a few zones.

2. Pressure vacuum breaker (PVB). The *PVB* is sometimes confused with the AVB because they look somewhat similar and both are installed above grade on nipples. There are, however, some important differences.

The PVB is installed on the mainline, on the upstream side of any control valves. The PVB is required to be a minimum of 12" above the highest head, or piping, in the sprinkler system. PVB's need to be tested by a certified tester when they are installed and annually thereafter. PVB's should also be tested whenever they are repaired or moved.

The PVB is similar to the AVB in that it has a poppet and a vent. When pressurized, the poppet seals the vent and permits water to flow through the device to the discharge side. If the incoming line pressure drops, this poppet will fall back, allowing water that is flowing in a reverse direction to spill out of the vent. Unlike the AVB, this poppet will not fall back into the throat of the device. Instead, a separate spring-loaded check valve shuts to

1-17. Pressure vacuum breaker (PVB).

prevent water from flowing back into the inlet piping.

PVB's have two small *test cocks* on the side of the body for the purpose of attaching a gauge to test the device. PVB's are typically sold as an assembly which includes a ball valve at the inlet side and a ball valve at the discharge side of the PVB body. These ball valves are for the purpose of isolating the body for testing, and should not be used as a substitute for a separate main shut-off valve. I will explain this in more detail in the design section of this book.

A trait shared by both AVB's and PVB's is that they are both installed above grade and they both have height requirements relative to the elevation of the sprinkler heads.

The height requirement can make placement impractical for some sites with substantial elevation differences. In those cases, the backflow device would need to be placed at the high point of the property, or might require nipples of an impractical height to elevate the device.

The fact that these devices are placed above grade may also raise concerns about aesthetics, exposure to physical damage, and in a cold climate, the question of winter freeze protection occurs. In one of these situations where there are negatives associated with either AVB's or a PVB, there are other options available.

3. Double check valve assembly (DCVA).

The *DCVA* consists of two spring-loaded one-way check valves that open in the direction of normal flow. If the incoming line pressure drops below the pressure on the discharge side of the device, the check valves slam shut and prevent water from flowing in a reverse direction. This will not only protect against backsiphonage, it will also protect against backpressure. Backpressure is the result of pressure being introduced into the upstream piping that exceeds the incoming water pressure, and so causes the water in the piping to flow in a reverse of normal direction. Sources of backpressure could be a pump in-line somewhere in the water system, or introducing compressed air into the sprinklers to winterize them.

Because it protects against backpressure, the DCVA is a good choice in freezing climates where compressed air is commonly used to purge water from the lines in the winter.

The DCVA may be installed underground in a vault, and so may be less obtrusive visually and less prone to vandalism, freezing and other physical damage than some other backflow devices.

The DCVA is also effective when installed at a lower elevation than the other system piping and can offer a solution for sites where PVB's or AVB's would not be practical.

The DCVA has four test cocks which are used to test the device. Like the PVB, the DCVA must be tested when installed and annually thereafter.

1-18. Double check valve assembly (DCVA).

The DCVA is typically sold as an assembly which includes a ball valve at the inlet side and a ball valve at the discharge side. These ball valves are for the purpose of isolating the device for testing, and are not a substitute for a separate main shut-off valve.

4. Reduced pressure device (RP). The *RP* is a device used in high-hazard situations. If you are incorporating a fertilizer injection system in your sprinkler system, or if you are pumping water from a well into a system that is also connected to a public water supply, you may be required to use an RP with your sprinkler system.

The RP device looks somewhat similar to the DCVA in that it also has two check valves. In addition, the RP has a relief valve situated between the two check valves that operates from a "zone of reduced pressure." When the pressure difference between the "zone" and the inlet pressure falls below a certain amount, the relief valve opens and dumps any water that may be flowing in a reverse direction. This provides protection even if the check valves were to become fouled.

Because of the relief valve, RP's must be installed in such a way that drainage is provided to permit the periodic dumping of water.

RP's can be installed above grade, or may be installed underground in a vault or otherwise indoors when acceptable drainage is provided for the relief valve.

LOCAL REQUIREMENTS

In conclusion, there are a variety of backflow devices available. You should always consult with your local water purveyor to make sure that you are in compliance with their requirements for the correct usage of the device. There may be instances where the local code is at a variance with even the manufactures recommendations for a particular device. Determine the appropriate backflow device for your system during the design phase of the project.

You want to consider both *which* device to install which is a water quality issue, and *how* to install it which can be a plumbing code issue. Be prepared to make a few phone calls and also ask your supplier for their suggestions.

Most places require a plumbing permit to install a backflow device.

1-19. Relative position of the components. Atmospheric vacuum breakers (AVB's) are installed on the discharge side of each control valve. Other backflow devices (PVB, DCVA or RP) are installed on the mainline on the upstream side of any control valves.

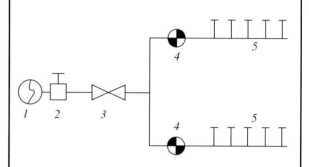

1. Water line to house.
2. Sprinkler main valve.
3. Backflow device- either PVB, DCVA or RP.
4. Control valves.
5. Sprinkler heads.

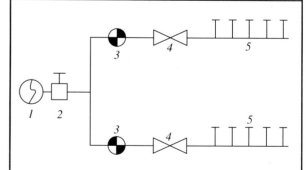

1. Water line to house.
2. Sprinkler main valve.
3. Control valves.
4. Backflow devices- AVB's.
5. Sprinkler heads.

Dos and dont's

Some common mistakes I have observed in do-it-yourself backflow installations:

1. The use of a combination control/anti-siphon valve for control valves in conjunction with a mainline backflow device, either a DCVA or a PVB. This is a redundancy. If the combination control/anti-siphon valves are correctly installed on each lateral line, then there is no need for a second backflow device on the mainline. Conversely, if a backflow device is installed on the mainline, the control valves do not need to have an additional anti-siphon feature.

2. AVB's installed too low or below grade: Most manufactures will recommend an AVB to be installed at least 6" above the highest head in the zone. I frequently see AVB's installed lower than this and worse, I have seen AVB's buried below grade in a valve box where the vent could become submerged or dirt could enter the vent.

3. AVB's installed on a mainline: Atmospheric vacuum breakers (AVB's) are not designed to operate under continuous pressure. They should be installed only on lateral lines, on the discharge side of the control valve.

4. Double check valve assemblies incorrectly installed: If a double check valve assembly is installed underground, it must be in a valve box that is large enough to permit access for testing and servicing. Adequate clearance must be available to remove the check valves from the valve body and to get gauge hoses attached to the test cocks. The serial number should also be readable. DCVA's should be installed on unions with enough clearance in the box to get a wrench on the union for removal. The valve box that contains the DCVA should contain **only** the DCVA. Do not clutter the installation by trying to cram control valves or anything else in the same box.

Pressure Regulator

A pressure regulator is installed on a sprinkler system when the incoming line pressure is so high that the working pressures of the sprinkler system are over what is recommended for the system components. Most sprinkler heads operate best between 25 and 75 PSI. Excessively high pressure is evidenced by spray that atomizes, or mists. The fine droplets of water are easily blown off target.

Most drip components operate at less than 30 PSI. Even normal household water pressures usually need to be reduced for drip irrigation.

Onward

You have now been introduced to the major components of a sprinkler system and some of the jargon that you will encounter.

Let's have some fun with this now, and move on to the next step which is to plan the sprinklers for your own yard.

1-20. A pressure reducer, suitable for installation on a mainline. The screw on top can be turned to regulate the discharge pressure.

Chapter 2

Design

In Design *I am going to show you, step-by-step, a logical sequence to develop a sprinkler plan for your home. You will learn how to determine how much water is available for your use, and how to size your sprinkler system to the available water. I will provide you with pointers on head selection and placement for maximum distribution uniformity. Valve selection and placement, pipe sizing, and zoning will also be covered. I will explain some basic rules of hydraulics in simple terms, and also show you how to calculate the performance of the sprinklers in your design so that you will know the plan is workable before the job is installed.*

The time you invest in your irrigation design will be time well spent. "Build in haste and repent in leisure." When you are done with this chapter, you could actually know what you are going to do before you start digging up the yard – what a concept!

In this section on design, we will examine the steps needed to design a sprinkler system. To do this, we will work through the process of planning a six-zone system for an average-sized city lot. This example will not be exactly the same as your own home. By working through this example, and understanding the concepts at work, you will then be able to custom design the sprinklers for your own home.

There are several things you will be doing to design the sprinkler system for your home that are the same things we would do to design the sprinklers for any home. I have drawn up a plan for you to look at. The house in the plan is not exactly the same as your house is, but it has the same elements that your house does, and we will want to deal with those elements in order to design the sprinkler system.

Here is a list of those elements, and the steps we can take to deal with them:

1. Site Analysis. *Site analysis* is just a way of saying that before we begin we want to know what we have to work with. In this case we want to know where the water is going to be coming from, how much water is available for our use and at what pressure. We also want to know about the soil and drainage. We want to note any significant elevation changes on the site. We also need to find out if we are going to be irrigating an existing landscape, or if the landscape is still in the planning stages.

Since the purpose of our sprinkler system is to irrigate the plants, it isn't difficult to see that we will need to have a landscape plan, so that we can design the sprinklers to be appropriate to the planting.

2. Head layout. After we know what the area to be irrigated looks like, we can then proceed by deciding what kind of sprinkler heads to use and where to place them for the best coverage.

In order to do this, we will explore some guidelines for head selection and head placement.

3. Zoning. *Zoning* means that we are now going to count the sprinkler heads that we placed in step 2, and add up the total amount of water required to water the entire yard. We will already know how much water is available because we will have determined that previously, in step 1. By dividing the total water required by the water available, we will know the maximum number of sprinkler heads that can be run at any one time. Then, we can factor in the other elements that concern us, such as portions of the yard that are in shade or full sunlight, plant communities with differing water requirements, slopes where runoff could occur and a few other things. When we are done with these steps, we will know how many *zones* will be needed in order to achieve the degree of control over the watering that we would like to have.

4. Point of Connection. Now we need to decide where the source point will be for the water that is to supply the sprinkler system, so I will discuss some parameters for making that decision.

5. Backflow prevention. We will select a type of backflow preventer that is appropriate to this job, and I will explain the rational behind the selection.

Having completed steps 1 through 5, the answers to some other questions will now become answerable: Valve locations, the number of valve boxes, the controller selection and location, and the low-voltage wiring path will all become clear now.

6. Pipe sizing. Now that the sprinklers and valves are placed (on paper anyway) we can decide where to run the piping necessary to get the water from where it is to where we want it to be. Once we know how much water each section of pipe is being asked to carry, we can calculate how big (diameter) each section of pipe needs to be, and how many lineal feet of each size of pipe are needed.

At this point we will be just about done with the design. I know this might seem like a lot of information the first time you do it, but I think you will find that by chunking it down into small steps and being methodical, it's really not too difficult. With some experience,

you could do the entire design in an hour or so. Being new, it will take you longer. But, whatever time you spend on your design will pay big dividends when you install the sprinklers. At the end of this section, I have provided a check list for your own design.
One last thing we can do is to use some standard tables I have provided, to determine if our proposed sprinkler system will operate within the limits of our water supply. By doing this, we can be confident that the sprinklers will work as planned when they are actually installed.

Let's go back and work through each of the steps I have just described to you, and apply each step to your own site. When we have completed this process, you will have an irrigation plan that is custom designed for your home.

1. Site Analysis

Water meter

The first thing we want to do is to figure out where your house gets its water from. Most homes in urban areas are serviced by a water district that maintains a network of piping under the city streets. Whenever a home is built, it is necessary to tap into one of these big water pipes and run a smaller service line to supply each residence. In most areas, water use is metered, and so a water meter is located on the property.

Where I live, in the Pacific Northwest, it's very typical for a city to maintain a utility easement along the front of every lot. Because of this, the water meters are almost always located close to the corner of the property near the street or sidewalk. Generally, the city owns everything up to and including the meter. The waterline from the meter to the house belongs to the homeowner.

In some cold parts of the country, where waterlines need to be placed very deep to protect them from freeze damage, you may find that the water meter is in the basement of the house.

There are some areas where an individual home may not have a meter.

If you get your water from a private well you would not necessarily need to meter your water use. And, some water systems have just not gotten around to installing meters yet.

The diagram 2-1 shows a typical city lot and water service arrangement. The water meter is conveniently located close to the property line in the utility easement near the street. The dashed line is the water service line that goes from the meter to the house and supplies the house with all of its water.

The first thing I want you to do is to locate your water meter. If you can't find it, call your local water purveyor (the city, water district, or utility company you get your water bill from) and ask them where it is. If you have a meter, they will know. Other bits of useful information you may be able to get from your water purveyor are the size of the water meter and the static water pressure. As you start to gather this information, get a worksheet started so that you can have all the information recorded in one location for easier reference later.

If you located your meter, lift up the hinged part of the meter cover and look in. You might be able to read the meter size stamped on the cover or printed on the dial. Be advised that meter pits may be home to spiders and snakes and so forth.

The most common residential water meter sizes are 5/8" or 3/4". One inch meters or larger are found much less frequently.

The significance of the meter size is that it tells us what the amount of water is that we can use in the design of our sprinkler system, provided that the water pressure and piping are adequate to accommodate that same flow. Most waterlines connecting a 5/8" or 3/4" meter with the house will be either 3/4" or 1" in diameter. The significance of the waterline size is that there is a limit to the amount of water that we want to flow through a pipe of a given diameter, just as there is a limit to the amount of water that we want to flow through

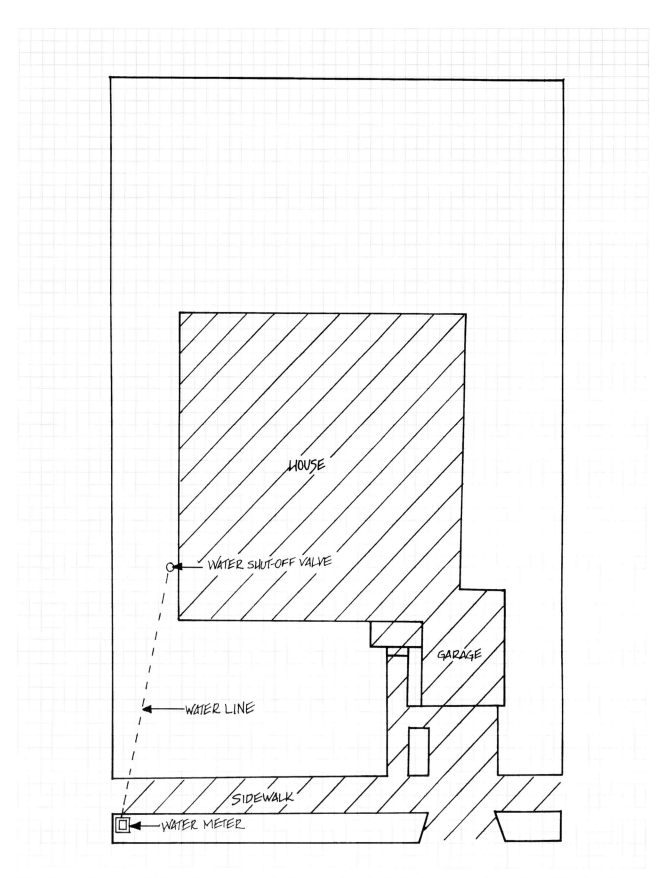

2-1. A base plan over which tracing paper can be placed to draw in the landscaping and sprinklers. The plan shows the house and permanent hard surfaces (driveway, walks). The water meter is in the parking strip between the sidewalk and the street. the dashed line from the meter to the house represents the water line. There is a shut-off valve near the foundation where the water line goes through the basement wall into the house.

a meter of a given size.

When we design the sprinkler system, we want to leave a buffer of 25% of the maximum safe flow of water through the meter. This buffer allows us to have some other household water use when the sprinklers are running.

The maximum recommended flows are as follows:

Meter size	Meter capacity	Useable for sprinklers
5/8" meter	20 GPM	15 GPM
3/4" meter	30 GPM	22 GPM
1" meter	50 GPM	37 GPM

The house in our exercise has a 5/8" meter, so I will consider 15 gallons per minute (GPM) to be the limit for the sprinkler system. If I had a 3/4" meter the limit would be 22 GPM and a 1" meter would give us 37 GPM to work with.

If you live in an area of high water pressure, you will be able to force more water than these recommended amounts through your meter, but I urge you not to do that. Exceeding these limits will cause damage to your meter and piping.

The meter size will tell you the maximum amount of water that may be available for your use Most of the time that amount will be available if the point-of-connection is made close to the meter. If the point-of-connection is made at any distance from the meter, than we must also consider the diameter and condition of the waterline up to that point-of-connection when determining the amount of water available for our design.

Not only do we want to flow an amount of water that is within the limits of our meter, we also want to stay within the limits of our pipe size. A 3/4" pipe should be used for flows of 10 GPM or less and a 1" pipe should be used for flows of 11 to 20 GPM. If we have a 5/8" meter we will have a potential for 15 GPM available for the sprinklers, but we will need a 1" pipe to use that full potential. If we have a 3/4" pipe with a 5/8" meter, than the limiting factor becomes the pipe and we need to limit our flow to 10 GPM.

A 5/8" meter will frequently have a 3/4" waterline going to the house. This means that if we want to use the full 15 GPM irrigation potential of that meter, we need to "tee" into the waterline very close to the meter so that we can run 1" pipe to our sprinkler valves.

Let's get a little technical background here so that I can explain this better.

Technical Stuff

My intention is to make this book as non-technical as possible. But, building a quality sprinkler system does require an understanding of some basic principles. Take a few minutes to read and understand these simple concepts presented here, and your sprinkler project will be more enjoyable and successful.

1. Static water pressure: *Static water pressure* is the force that water in a closed system of pipes exerts against the walls of the pipes. Static pressure is measured pounds per square inch (PSI) that is being exerted against the walls of the pipe. To find out what your static water pressure is, you can call your water purveyor and they should be able to tell you the static pressure at the meter. You can also find out your static water pressure by attaching a pressure gauge to an outside hose bib and turning the water on. When you do that, make sure that no other water is running anywhere in the house or you will not have an accurate reading.

2. Working water pressure: *Working water pressure*, also called *dynamic* pressure, is the water pressure in the pipes when the water is running. Working pressure is also measured in PSI. The difference between static pressure and working pressure is that static pressure is equal everywhere in the system of pipes. If you have a static pressure of 50 PSI at the water meter you will have a static pressure of 50 PSI at the hose bib by the house, even though the hose bib might be 100' away from the meter. Remember, in a static condition, no water is moving in the pipes.

Just suppose, though, that we open up the hose bib and let the water flow. Now that the water is moving in the pipe, there will be pressure loss from friction along the entire length of the pipe. The 50 PSI that we started out with at the meter might be down to 45 PSI, or some other pressure, by the time it comes out of the hose bib. Just how much pressure will be lost to friction depends on several variables. The main factors that affect how much pressure will be lost to friction are the condition of the pipe walls (rough or smooth) and the *velocity* of the water as it travels in the pipe.

3. Velocity. The *velocity*, or speed, that the water travels is a function of how much water is trying to get through the pipe, relative to the diameter of the pipe.

For example, have you ever been watering with a garden hose and partially blocked the stream of water as it left the end of the hose by covering it with your thumb? Just about everyone has done that. And, most likely you noticed that you got a more powerful stream of water when you did that, and that the stream of water seemed to travel further and knock the dirt off of your car or sidewalk better. Most people would say that they got more "pressure" when that happened. Actually, what happened is you got less *pressure* but more *velocity*. When you put your thumb over the end of the hose, you made the opening smaller. In order for the same amount of water to get out, it had to speed up. The water pressure, right at that point, was actually reduced because your thumb introduced more friction into the system. But, due to the smaller opening, the velocity was increased. That is why the water squirted a greater distance and knocked the dirt off better.

As you will see, the significance of this for our sprinkler design is that we want the water to travel in the sprinkler pipes at a certain velocity, and we can control this velocity by the diameter of the pipe we use.

4. Flow rate. The *flow rate* is the amount of water moving past a fixed point in a given amount of time. Flow rates are measured in gallons-per-minute (GPM) in conventional sprinkler systems and gallons-per-hour (GPH) in low-volume (drip) systems.

Flow is the volume, or amount, of water we have to work with. It is important to distinguish flow from pressure, which is the force pushing the water. The reason why this is important is because it is possible to have high pressure with low flow. You might put a pressure gauge on a hose bib and get a high static pressure reading and yet, when you remove the gauge and let the water run, only a weak stream of water comes out. This could be the result of an obstruction or a kink in the water line, or corrosion in the water line which has restricted the flow.

When we design our sprinkler system we will want to consider *static pressure, working pressure, flow,* and *velocity*. All four of these are interrelated and changing one may affect the others. Don't be too concerned about the terminology. I just want you to be introduced to these ideas so that you can understand how your sprinklers work.

O.K., you have now located your water meter and you have made a note of the meter size and your static water pressure.

The next thing we want to do is to get an idea of the flow that is available. To do this you will need a 5 gallon bucket and a watch with a second hand or a stopwatch. Place the bucket under the hose bib that is closest to the meter and turn the water on all the way. See how long it takes to fill the 5 gallon bucket. Once you know how many seconds it takes to get 5 gallons of water in the bucket, you can calculate how many gallons per minute are flowing. For example, if it takes 20 seconds to fill the 5 gallon bucket, then you could fill 3 buckets in a minute, or you have a flow of 15 GPM.

Formula for calculating GPM

$$\frac{60}{T} \times 5 = GPM$$

Where T = the time in seconds that it takes to fill a 5 gallon bucket

---

Using this method to determine your flow will give you a very conservative number. Chances are that the hose bib is a considerable distance from the meter and, unless you are in a new home, the water line may have some corrosion. Even in new homes, I have occasionally seen brand new copper service lines that had dents and kinks that would restrict the flow. Additionally, the water may travel through even smaller pipes in your basement before coming back out through the foundation and ending in a hose bib. Regardless, record this flow number now and we may change it later when we tap into the service line. One of the reasons why I like to connect the irrigation mainline as close to the water meter as possible is because then we will be using new plastic pipe, with good flow characteristics, that is of a large enough diameter to let us use an amount of water that is at the maximum safe capacity of the meter.

Locate the Waterline

Looking back again to Figure 2-1, you can see that there is a shut-off valve on the side of the house where the waterline would go through the foundation to connect with the inside piping.

Because the waterline starts at the meter and runs to the shut-off valve, we have two places where we will be able to locate that waterline by digging down- at the meter and at the valve.

The waterline may run in a straight line between those two points but don't bet the farm on it. Sometimes these pipes take some interesting detours that only the plumber who installed it could tell you about. There are devices you can rent to track the path of the pipe, but you probably won't have any reason to do that. Most of the time you will want to connect the sprinklers as close to the meter as possible. If the waterline to the house is in good condition, and if the diameter of the waterline is adequate to supply the sprinkler system, then you could also tap into it at a location that was further from the meter. In that case, the second most likely place to dig down and locate the waterline at this house would be at the shut-off valve by the foundation. That is simply because the pipe is easier to locate there because the presence of the valve indicates the location of the pipe.

Your local building code will specify at what depth you can expect to find a waterline. In Portland, Oregon most waterlines, by code, are 2 feet below grade. In some colder climates it could be much deeper.

In newer homes, the waterline will be plastic or copper. In older homes, the waterlines were made of galvanized steel. If you have recently moved into an older home, it is possible that the previous owner may have, by now, replaced the original steel pipe with copper or plastic. This is because the steel does corrode and need replacement eventually. You may not have any way of knowing for sure what kind of waterline you have until you dig it up and can look at it.

Basement connection

You may live in a home with a full basement. If so, it is possible that you can locate the waterline pipe where it comes in through the basement wall to feed the household network of pipes. If that pipe is of a large enough diameter and is in good condition, then this is another point at which you may have the option of connecting the sprinkler system.

If you do connect in the basement, the most common way would be to install a gate and waste valve on the basement waterline and then run the sprinkler mainline from that point back through the basement wall below grade, and then proceed with the sprinkler system from there. Doing this requires that you drill through the basement wall with a roto-hammer.

Wells

Just suppose you get your water from a private well—does any of the above apply to you? Yes. All the design principles we will talk about apply equally, regardless of your water source. If you have a well you have a few more things to consider.

You need to design your sprinklers to operate within the limits of your pump and water source. Keep in mind that your well is going to produce a given GPM, and changing a pump size will not give you more water. The pump will move the water that is there, but it can't make water. You may also have a holding tank or a pond you pump into. There are lots of variables, and I don't know what your individual situation is. You may want to have a well company do a flow test for you. You may have good flow for a few minutes, but the sprinklers may be running for several hours if you have a large place, and you will want to know what kind of a flow is sustainable.

When you are getting water from a well, or pumping from a river or pond, there is a good chance that you will want some filtration on your system to keep contaminates from fouling the sprinkler parts, especially electric control valves and gear driven sprinklers. Check with your local supplier for a recommendation on "dirty water" valves if you have this situation. Impact heads will resist clogging the best if the water is not clean.

Backflow prevention is just as necessary on a well as it is on city water. The environmental laws that protect aquifers are very strict.

The bottom line is, you need to come up with some idea of your pressure and flow to design the sprinkler system. The less information you have, the more hit and miss your design will be.

ELEVATION

You want to make a note of any significant elevation changes on your property. Water will lose .433 PSI for every vertical foot that it travels uphill and gain .433 PSI for every vertical foot that it travels downhill.

Just suppose you live on a hillside lot and your water meter is at the street with a static pressure of 50 PSI. And just suppose that the elevation difference from the meter to the highest point in your backyard is 20 feet. That means that water piped to the backyard would lose over 8 PSI just in elevation loss alone, even before we subtract the other pressure loss from friction in the pipe and valves and so on. When we design our sprinklers, we cannot ignore the effects of elevation.

Another reason why we want to pay attention to elevation is because we want to anticipate drainage patterns and we want to, as much as possible, try to predict how much water can be applied to a slope before the water starts to run off.

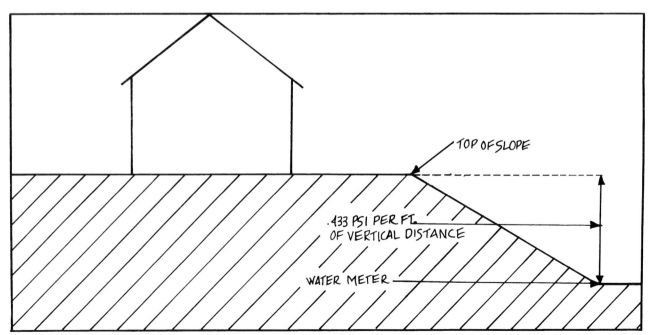

2-2. *Water loses .433 PSI of pressure for each vertical foot of elevation gain. Water also gains .433 PSI of pressure for each vertical foot of elevation loss.*

If you think about it a little bit, I think you will see that this is going to depend on the steepness of the slope in combination with the porosity of the soil. If a soil is very tight, like a clay soil, then very little water can be applied before runoff occurs. If a soil is more course, like a sandy loam, then the water will percolate into the ground better, and less runoff will occur.

There are several things we can do with the sprinkler design to compensate for steep slopes.

Choose sprinklers with a low precipitation rate so that water is applied slowly. That way, the water has a chance to soak in. Applying water at a rate that is faster then the soil can absorb will lead to runoff.

Another thing we can do is to have the sprinklers on the top part of the slope run on a different control valve than the sprinklers on the lower part of the slope. By doing that, we can water the bottom of the slope for less time, knowing that it is receiving additional moisture in the form of runoff from the upper slope.

Landscape plan

Since the purpose of our sprinkler system is to water the plants, we need to have a landscape plan before we can design a sprinkler system. If you are installing sprinklers at a site that is already planted, and no changes in the planting are anticipated, then you can move right along to the sprinkler head layout.

If you are irrigating a yard where there is no landscape yet, or if you will be remodeling an existing yard, then you need to make some decisions about plant placement before irrigating.

You do not need a very elaborate landscape plan for the sprinklers. What you do need is at least to decide what will be the areas of lawn, shrubs and groundcovers. The reason for this is because, at the very least, we want to be able to water the lawn and shrub areas separately. That is because the water requirements will be very different for a lawn as compared to the water needs of established shrubs.

Just suppose your landscape plans are still evolving and you haven't picked out all the plants yet. That's fine, but at least know where the bedlines will be and any major landscape features before laying out the sprinklers.

You do not need to make a big investment in drafting supplies. Just have a ruler or an architect's scale, some graph paper, a pencil, and a tape measure. Pick a scale that gets your entire lot on one piece of paper. One thing you can do is to get the lot, house and permanent paved surfaces drawn, and then use that as a base sheet over which you can overlay tracing paper to try out different landscape designs. When you get the landscape you want, then you can use that as your base and use a tracing paper overlay to try out different sprinkler layouts.

You can also photocopy the base sheets and use the copies to try out your ideas instead of tracing paper.

The first thing I did was to get some 11 x 17" graph paper. The paper has 1/8" grids which makes it a convenient size because the lot we are working with in this example is 105' x 64'. By making 1/8" = 1', the lot just fits on one sheet of paper, with a little room left over for notes in the margins.

The next thing I did was to draw in the house, garage, driveway and sidewalks.

I also marked the location of the water meter and the main water shut-off valve at the house.

Take a look at Figure 2-1. The dashed line between the meter and the valve represents a guess of the waterline location. At this point in time, I really don't know if that is the actual location of the pipe, it's just the most direct path from the water meter to the shut-off valve.

Once I had the permanent features drawn on the graph paper, I made a few photocopies. Now I can use the copies as a base plan to try out various landscape and irrigation ideas. I could accomplish the same thing by making a base plan and then using tracing paper as an overlay to try out different ideas. To do it that

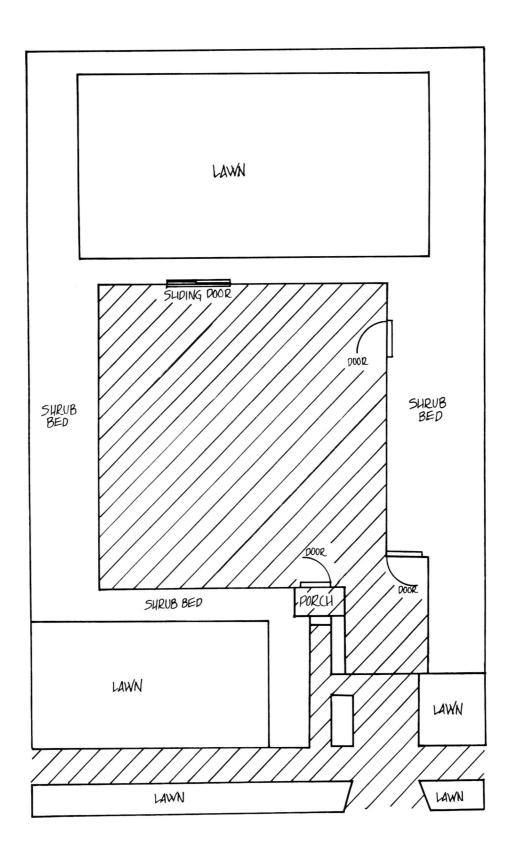

2-3. This is the minimum landscape plan you will need before you can design the sprinklers. This plan defines which areas are lawn and which areas are shrub bed. It also shows the location of the house, walks and driveway.

way, put the tracing paper over the base plan and draw in the plants and other landscape features. To try different ideas, take off that sheet of tracing paper and put a fresh sheet over the same base plan. That way, you can invent as many different landscapes as you want, without redrawing the lot lines and permanent structures every time. Once you have the landscape plan finished, you can trace in the base plan on the same sheet. Now you have a complete plan. The next step is to take this completed landscape plan and overlay it with tracing paper again, on which you can now design the sprinkler system. This way, you can try out as many sprinkler lay-outs as you need to, without having to redraw the entire landscape plan every time.

It's not absolutely necessary for you to have a plan on paper. It may be that you can visualize the landscape better if you physically lay-out the actual site. I like to paint bed lines with marking paint, but you could also lay a garden hose or rope on the ground to represent where the edges of the beds will be. Stakes can be used to represent specific plant locations.

Take a look at the landscape plan for the house in our example. It is not elaborate, but it is accurately drawn to scale. It shows the position of the house on the lot, all the hardscape surfaces and the separate lawn and shrub areas.

Figure 2-3 shows the absolute minimum you will need for a landscape plan, either on paper or physically laid out on the actual site, before you can start designing the sprinklers. In this situation, without any more specific information, I would lay-out the sprinklers in the lawn areas for head to head coverage. I would also layout the sprinklers in the bed areas to cover the entire bed area. The amount of coverage will be the same, but I will want different heads in the shrub beds then are covering the lawn areas because I know that I will want the lawn and shrubs to be on different valves.

Laying out sprinkler heads to water the entire bed area will be acceptable if the shrub beds are, in fact, going to be densely planted. If the shrub beds end up being sparsely planted, however, irrigating for full coverage will mean that we waste water and encourage weed growth by watering too much empty ground between the plants. If we had a more detailed planting plan than this, we could do a better job of designing the sprinkler system.

Figure 2-4 shows a more detailed planting plan for the shrub beds. Some shrubs are designated only as "small shrub" or "medium shrub" because at this stage of the planning the final selection has not yet been made. On your own plan, fill in as much information as you know at the time the plan is drawn. Even having some specific plant locations marked gives us a much better idea of where to place the sprinkler heads than we had before. The shrub areas that are not planted will be covered with a bark mulch. We are going to use Figure 2-4 as the landscape plan to design our sprinkler system. We will use all conventional irrigation. In the next chapter, when we look at low-volume irrigation, we will irrigate this same site with drip irrigation so that you can compare the two methods.

There is computer software available for both landscape and irrigation design. I'm not going to make any recommendations in this regard because there are so many variables in software programs, hardware you may own to run it, and your own computer skills.

If you are not already familiar with the specific application software, then consider your time to install it and learn it. You would have to be the judge of whether it is worth it for you to get geared up to design one sprinkler system, and then how do you evaluate if the end product is really a worthwhile design?

SPRINKLER HEAD SELECTION AND LAYOUT

In Chapter One we covered the various types of sprinkler heads that you could choose from. Let's now take a look at some general information regarding the correct placement of sprinkler heads. Here are some useful terms and concepts for you to be acquainted with:

1. Discharge rate: *Discharge rate* is the rate

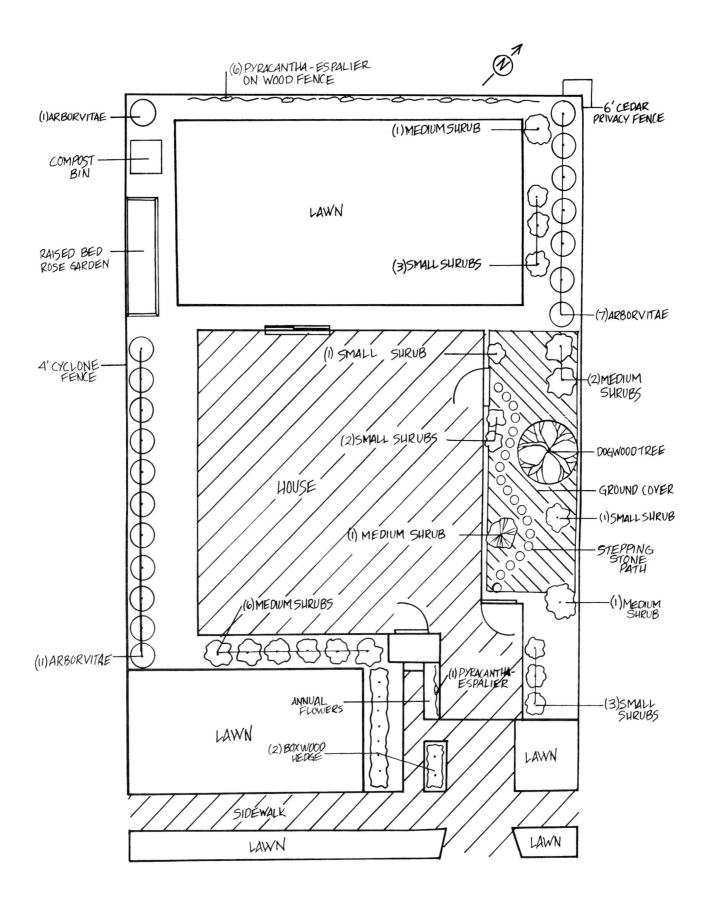

2-4. This is a progression of the landscape plan, showing more planting detail. Having the plants located gives us a much better idea of how to place the shrub heads than we had before.

at which water is discharged from a particular nozzle at a given water pressure. Discharge rates for conventional sprinklers are given in gallons- per-minute (GPM). For example, I might have a quarter-circle nozzle with a 15' radius of spray that discharges .33 GPM. A nozzle from a different manufacturer might spray the same radius and pattern but discharge .5 GPM. You will find the discharge rates and other data for nozzles available in the manufactures literature where you buy your sprinklers.

We want to know the discharge rate for each nozzle in our sprinkler system, because by adding those up, we can find out how much water is needed to run the sprinklers. Additionally, we want the discharge rates of the nozzles on any given zone to be such that we have matched *precipitation rates.*

2. Precipitation rate: The *precipitation rate* is the rate at which the water discharged from a sprinkler, or group of sprinklers, lands on the ground or plants. It (precipitation rate) is a measure of how many inches of precipitation are falling per hour, just like when your local weather announcer tells you that an inch of rain has fallen in the last 24 hours.

If you were to run a single sprinkler head by itself, and put several catch cans on the ground at different locations within the radius of spray, you would observe that when you were done watering, the cans held different amounts of water. The water in the cans would represent the precipitation rate at each location.

Sprinklers do not discharge water evenly within their entire radius of spray. When we are placing sprinklers, we almost always place them so that every square foot of ground is being watered by more than one head. This is to compensate for the lack of uniformity in coverage from any one sprinkler nozzle. By having sprinklers spray "head to head", we achieve the most even precipitation rate throughout the entire zone.

3. Matched precipitation rates: This is a concept of selecting sprinkler nozzles by *discharge rate,* so that when they are combined in a zone the net effect is the most uniform *precipitation rate.*

When manufactures offer nozzles in sets that are designed for matched precipitation, it means that if the 1/4 circle nozzle puts out 1 GPM, then the 1/2 circle nozzle will put out 2 GPM and the full-circle nozzle will put out 4 GPM, and so on. Spray head nozzles are offered in matched precipitation rate sets by the radius, or distance, of spray. For example, in a 10' radius set the different spray patterns will have matched precipitation rates relative to the other 10' nozzles.

However, if you use rotor or impact type heads, you will need to select nozzles yourself to match precipitation rates. This is easy. If a rotor head has a 1/4 circle throw and a radius of 25', then pick a nozzle for your 1/2 circle heads (with the same radius) that has approximately twice the discharge rate as the 1/4 circle.

What we are trying to do is discharge the same amount of water per square foot. So if a sprinkler is covering twice the square footage, it needs to discharge twice the amount of water, relative to another sprinkler watering on the same zone.

4. Distribution uniformity. *Distribution uniformity (DU)* is a measurement of how uniform the precipitation rate is throughout a given zone. Matched precipitation rate nozzles, correct operating pressures, and head to head spacing will result in a sprinkler system with a high degree of distribution uniformity. The higher the distribution uniformity of a zone, the more efficient the scheduling can be. When a sprinkler system has a low distribution uniformity, especially in lawn areas, there will always be parts of the lawn that are over-watered or under-watered. What usually happens is that a zone will run for a certain amount of time and most of the area covered will be adequately watered, with the exception of a few dry spots. In an attempt to green up the dry spots, the owner will increase the watering time. The dry spots may eventually catch enough run-off to green up, but the price will be that now the rest of the lawn is over-watered. A professional landscape irrigation water auditor has ways of measuring distribution uniformity so that the efficiency of a sprinkler system can be expressed as a percentage and actually rated. This is more technical than we need to get for our purposes. The basic idea for you to have when you design your sprinklers is that even, uniform coverage is good, because that will enable you to keep the plants healthy with minimal water use. Uneven coverage is to be avoided as much as possible, because that will lead to some parts of the yard being over-watered, and some parts of the yard being too dry, while using excessive amounts of water.

5. Head spacing: *Head spacing* is the distance apart that we put the sprinklers. In a lawn area, it is very important to have *head to head* spacing. That means that a head with a 12' radius nozzle will be placed 12' away from the head next to it, so that the spray from each head reaches all the way to the adjacent head. In an area where there is a prevailing wind, you will space the heads even closer, so that there is enough overlap to compensate for the distortion of the spray pattern caused by the wind.

In a shrub bed, head to head coverage is not necessary. Shrub heads can be spaced to achieve 70-80% coverage. The reason for this is because the shrubs only need 70% of their root area watered to thrive. Shrubs send out roots to collect water, and so it is not necessary to wet every square inch of bed area. Wetting the foliage on shrubbery doesn't do anything except waste water to evaporation.

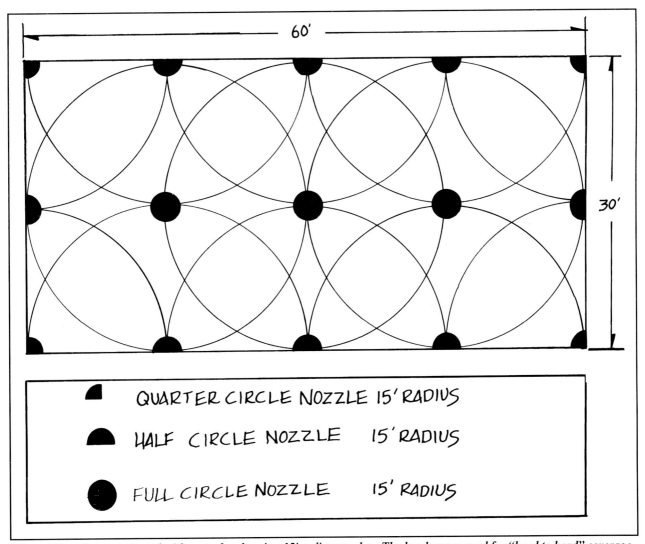

2-5. A 30' x 60' area irrigated with spray heads using 15' radius nozzles. The heads are spaced for "head to head" coverage.

Once in a while it's good to spray the dust off the leaves, otherwise it is only the roots that need the water.

Low-volume (drip) irrigation can be very effective watering shrub beds. Chapter Three in this book deals with low-volume watering.

6. Overspray: *Overspray* or *overthrow* is the water that misses the target area and wets an adjacent surface. Using sprinklers that have too large a radius of spray, or the wrong pattern for a given area, contributes to overspray. Excessively high water pressure can also contribute to overspray because the high pressure will cause the water droplets to mist when they leave the nozzle, and the fine droplets drift off target.

Overthrow onto non-landscape surfaces is always a waste of water, and can cause water spots on windows and stains on walls and fences.

Sometimes, overthrow is used deliberately as a way to cut corners on an installation, as when someone may install lawn sprinklers, and try to water the adjacent shrubs with overspray from the lawn heads, rather than installing additional shrub heads.

7. Head selection: We will choose the head that gives us a nozzle choice adequate to water the area in question with the fewest number of heads. For example, if we have a rectangular lawn that is 15' wide by 30' long, we will use spray heads with 15' radius nozzles. If we have a rectangular lawn that is 30' wide by 60' long, we could use spray heads but, chances are we would select rotor heads and nozzle and adjust them for a 30' radius.

Odd shaped lawns with both wide and narrow areas are challenging. Sometimes you will want to use spray heads for the narrow

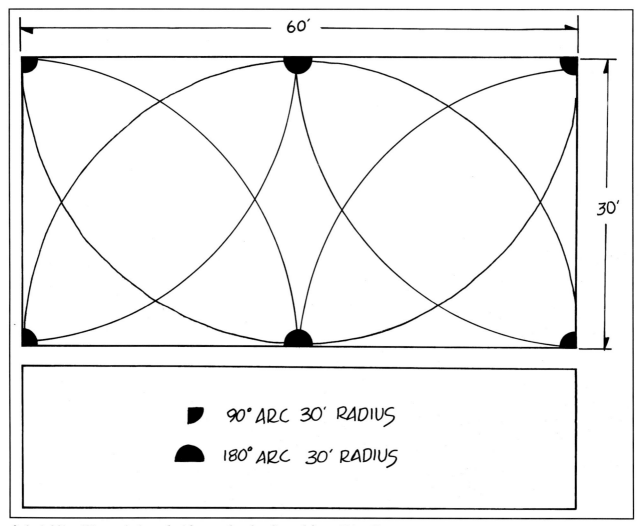

2-6. A 30' x 60' area irrigated with rotor heads adjusted for a 30' radius.

parts and rotors for the more open areas. If you do this, it is very important that the spray heads and rotors be on different valves. This is because the spray heads will have a much shorter run time than will the rotors. You should never put unlike head types on the same valve.

Lawns with curved edges and complicated shapes are usually best handled by using smaller radius nozzles. This gives you better control of the water and minimizes overspray.

Now that you have been introduced to some basic concepts and terminology, we can start working on the actual design.

STEP 1

HEAD LAYOUT

Lay a sheet of tracing paper over your completed landscape plan. Have on hand a catalogue of the sprinklers you plan on using. The catalogue will be a reference as to the sprinkler heads and nozzles you will be selecting. To layout the sprinkler heads, you will need to know the shapes and radius of the spray patterns that are available to you, as well as the amount of water required to run each nozzle, and the pressure requirements of the heads. If you are familiar with the specifications of a particular product line, this is very easy. If you have never done this before, it will naturally be a little bewildering at first. I suggest that you first decide what brand of sprinklers you are going to use, and then spend some time looking through that manufactures catalogue. Just being aware of what's in the catalogue will help you make choices about sprinkler selection.

Figure 2-10 is an example of the sprinkler heads I laid out for this site on some tracing paper. The number to the right of each head on the plan indicates the maximum radius (in

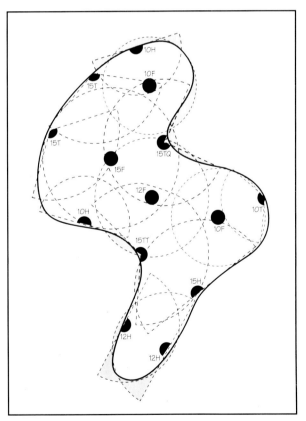

2-7. Free-form lawns with curved edges can be challenging to water because it is difficult to avoid some overspray. This one is irrigated with twelve spray heads, nozzled as indicated on the drawing. the shaded areas represent overspray.

2-8. This is the same lawn as Figure 2-7. This time it is irrigated with 14 spray heads, nozzled as indicated. Notice that by using two more heads and smaller nozzles than were used in the first example, we have eliminated most of the overspray.

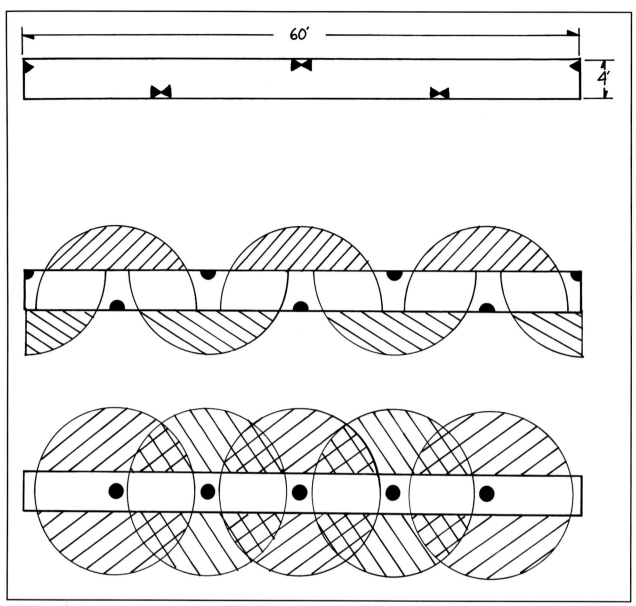

2-9. Top: A 4' x 60' area correctly irrigated with two 15' end strip and three 15' side strip nozzles. This provides 100% coverage with no overspray. Middle: The same area with quarter-circle and half-circle nozzles. The area is watered, but there is too much overspray, as indicated by the shaded areas. Bottom: Full-circle heads will water the area but, again, the excessive overspray is not acceptable.

feet) of the selected nozzle. The letter indicates the shape of the spray pattern. The number directly underneath each nozzle designation is the gallonage (in GPM) that the nozzle will discharge at it's working pressure.

Here is what I was thinking as I laid out each area:

1. Front parking strip (narrow lawn area between street and sidewalk): Because the shape of the lawn is a narrow rectangle, this is a good place for side strip and end strip nozzles. We will use heads with a 4" pop-up height in all the lawn areas to get clearance over the top of the grass. In the small lawn on the north side of the driveway, we can use quarter circle nozzles in the corners, and use the adjusting screws on the nozzles to reduce overspray.

2. In the front and back lawns we will use adjustable rotors with a 4" pop-up height. The rotors in front will be adjusted to a radius of 17' and the rotors in back will have a radius of 25'. Notice that we have selected nozzles for the rotors so that the half-circle heads discharge double the GPM that the quarter-

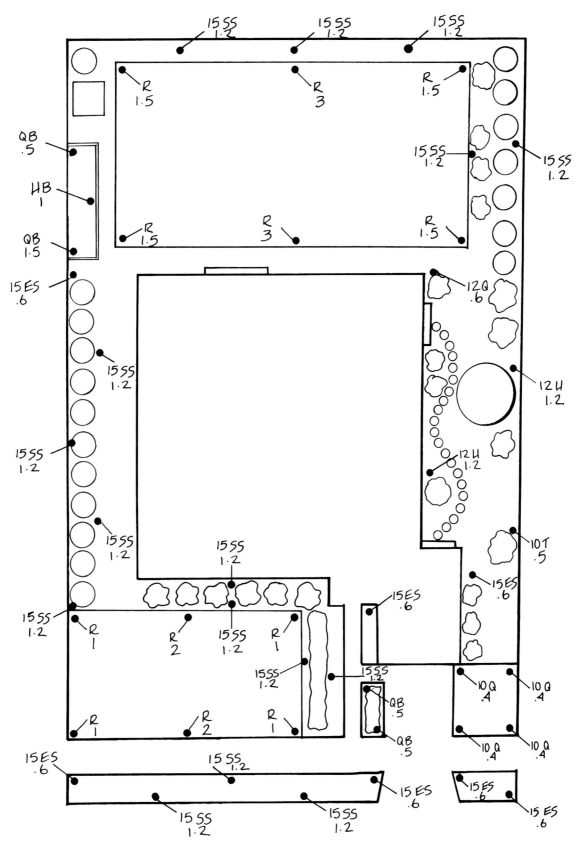

2-10. This is how you can draw in the sprinkler heads on a piece of tracing paper laid over the landscape plan. Just put a dot where the head goes, and make a note near it that will tell you the nozzle size and type, and the discharge rate. For example "10Q" tells me I'm looking at a quarter-circle nozzle with a 10' radius of spray. Seeing ".4" written near that tells me that the 10Q nozzle has a discharge rate of .4 GPM.

circle heads discharge. That gives us uniform coverage, with a head covering twice the area delivering twice the water relative to the other heads on the same zone.

3. The shrub areas are laid out to achieve coverage with a minimum of overspray. Shrub areas often contain odd nooks and crannies that can be challenging to irrigate. Match the nozzles to the shape and size of the area as closely as you can, and use the adjusting screw on the nozzles to fine tune the coverage.

Some manufactures supply "pressure compensating" filter screens and/or nozzles which are very helpful when trying to fine tune small areas. Pressure compensating screens will allow you to "throttle down" a nozzle more than you would be able to with a regular screen.

In this plan, we will use 4" pop-ups in the shrub areas along the arborvitae hedges and also along the back fence where the pyracantha is espaliered. In the raised bed planter, I will put the stream bubblers on nipples with a shrub adapter. We will have the nozzles located just an inch or two above the top of the soil, so that the fingers of water don't get the foliage of the roses wet.

In the shrub area with ground cover I will use 6" pop-up spray heads. I will also use a 6" pop-up for the end strip nozzle on the north side of the garage, to get the spray up and over the shrubs.

STEP 2

ADD UP GALLONAGE

The next step is to add up the total gallons per minute that will be required to run all the sprinkler heads. To make it easier, I chunked it down into four areas that seemed like natural divisions to me. These are not really zones yet, I'm just making some obvious distinctions to make it easier to organize and gather some information: (1) The spray heads in the front lawns (2) The rotor heads in the front lawn (3) The rotor heads in the back lawn and (4) The spray heads in the shrub beds.

What I did was list each nozzle and the GPM that was given in the manufactures specifications in their catalogue (which I got from the salesperson where I will buy my sprinkler parts).

For example, the first front lawn spray head nozzle listed is a "15 ES" which is an end strip nozzle with a 15' throw. In The "GPM" column to the right of the nozzle is listed the discharge rate for that nozzle, which in this case is .6 GPM. To the right of that, in the "Qty" column is the quantity, or number of these nozzles found in the front lawn, and we see that we have counted four. Four heads at .6 GPM is 2.4 GPM which is the number in the far right column. In like manner, we add up the total GPM of all the sprinkler heads in our design.

1. Front lawn spray heads:

Nozzle	GPM	Qty	Total GPM
15 ES	.6	4	2.4
15SS	1.2	3	3.6
10Q	.4	4	1.6

Total gallonage-front lawn spray heads .... 7.6 GPM

2. Front lawn rotors:

R	1	4	4.0
R	2	2	4.0

Total gallonage-front lawn rotors .... 8.0 GPM

3. Back lawn rotors:

R	1.5	4	6.0
R	3	2	6.0

Total gallonage-back lawn rotors .... 12.0 GPM

4. Shrub bed spray heads:

SS	1.2	12	14.4
ES	.6	4	2.4
QB	.5	4	2.0
HB	1	1	1.0
12Q	.6	1	.6
12H	1.2	2	2.4
10T	.5	1	.5

Total gallonage-shrub bed spray heads .... 23.3 GPM
Total combined gallonage of all sprinklers ... 50.9 GPM

Now we can use this information to assist us in deciding how to zone the sprinkler system.

STEP 3

ZONING

You will recall the beginning of the chapter when we gathered some site data, including the water meter size, water pressure and flow rate. At the house in our example, we have a 5/8" meter, 55 PSI static pressure, a 1" copper waterline to the house, and 12 GPM flow from the hose bib. The maximum safe flow through a 5/8 meter, we know is 15 GPM.

If we add up the total gallonage demanded by our sprinkler head layout (which we just completed) we find out that our sprinkler system will use 50.9 GPM. But, we can only use a maximum flow of 15 GPM, because that is the limit of our meter capacity.

You can see that, to make this work, we will need to divide the total demand of 51 GPM (I rounded off) into several *zones*. Each zone will be a group of sprinkler heads that is connected to a single control valve. Each zone will demand a total GPM that is less than our designated maximum flow of 15 GPM. When we water, we will run one zone at a time so that we never exceed our maximum safe flow.

Dividing 51 GPM (total demand) by 15 GPM (maximum available water) gives us 3.4. That means that we will need 4 zones to stay within the limits of our water source.

In other words, at this point in the design phase we know that, depending on where we ultimately decide to tap into our waterline, we are going to have either 12 or 15 GPM to work with. We also know that the sprinkler heads we just placed on paper require a total of 51 GPM. Therefore we can anticipate splitting the sprinkler heads into at least 4 different *circuits*, or *zones* to stay within the limits of our water source.

In summary, do these steps for your own home:
1. Determine the available water.
2. Lay-out the sprinkler heads.
3. Add up the total gallonage demanded by all the sprinklers.
4. Divide the total demand in GPM by the available water in GPM.
5. The number you get will be the minimum number of zones, or circuits, needed to run your sprinklers.

In addition to staying within the limits of our water source, we also want to make some other distinctions about how we zone our sprinklers. The reason for this is because every head that is run from the same control valve runs at the same time. This may seem obvious, but we need to consider that different plant types have different water requirements. And, different exposures (sun, shade, wind) will change water requirements at different places in the yard. For example, a narrow side yard that is shaded most of the day, and is planted with drought tolerant shrubs, will have a very different watering schedule than a lawn planted on a sunny exposure. You would not want the sprinkler heads for these two areas connected to the same control valve. If they were (connected to the same control valve) than you would find that to water the lawn adequately would require over-watering the shrubs, and that if the water, in this situation, were cut back to an amount that was appropriate for the shrubs, then the lawn would be under-watered.

To avoid this, we want to examine our site and identify like plant groups and microclimates so that we can consider plant watering needs, as well as available water, when we are deciding how to zone our sprinklers.

Another item to consider is the sprinkler head types and precipitation rates. We don't want sprinkler heads that aren't matched to be running together on the same valve. For example, spray heads should never share a valve with rotors.

Looking back at our landscape plan and sprinkler head layout, I see what I would identify as five distinct areas that could be classified as different hydrozones. When I add up the gallonage for each zone I get:

1. Front lawn spray heads. I identify this as a zone because it is the same plant (turf grass), same exposure (SE) with additional reflected heat from the street and sidewalk, and same head type. This zone has a demand of 7.6 GPM. I'm comfortable running all these heads on the same valve because the entire zone has the same water needs and is covered by heads with very similar precipitation rates. Also, the 7.6 GPM is well within our available water supply of at least 12 GPM. For similar reasons, I have also identified these as hydrozones:

2. Front lawn, rotors. Eight GPM. (Same head type, same plant, same exposure).

3. Back lawn, rotors. Twelve GPM. (Same head type, same plant, same exposure).

4. Shrubs in the shade on the NE side of house. Just over 5 GPM. (Same head type, similar plants, same exposure).

5. The last hydrozone is the shrubs in the sun. (Same head type, similar plants, same exposure). The heads watering that area add up to 18 GPM. Because that exceeds our maximum safe flow, I will divide that into two zones of approximately 9 GPM each.

That will give us a total of 6 zones required to run our sprinkler system. We arrived at that conclusion by taking into consideration the amount of water available, the types of sprinkler heads, the various plant groups, and the different exposures of the site.

Now that we know how many zones we need, we can make some additional distinctions.

STEP 5

DETERMINE POINT-OF-CONNECTION

Because our largest zone requires 12 GPM, we will be able to make our point of connection near the shut-off valve by the foundation. The meter, we know, has a maximum safe flow for sprinklers of 15 GPM. That is 75% of the meter's full capacity of 20 GPM. The 1" copper waterline to the house has a capacity of 20 GPM. In theory then, we could tap this waterline anywhere along it's length from the meter to the house and have access to at least 15 GPM for our sprinklers. At 15 GPM, we are not exceeding the capacity of either a 5/8" meter or a 1" copper pipe. But, just suppose we only had a 3/4" waterline to the house. In that situation, I would recommend putting the point-of-connection close to the meter so that we could increase the pipe size to our sprinklers to 1".

STEP 6

DETERMINE APPROPRIATE BACKFLOW PREVENTION

For backflow protection we are going to use a double check valve assembly (DCVA). If we used atmospheric vacuum breakers (AVB's) in this design, we would need six of them (one for each zone) which would be an eyesore and not cost effective. We decided against a pressure vacuum breaker (PVB) because it would also have to be installed above grade and be visible. Additionally, we live in a freezing climate, and we want as much of the sprinklers below ground as possible. We also want backpressure protection when we winterize with compressed air, which is another good reason to use a DCVA. We won't use a reduced pressure principle (RP) device because in this situation it is more protection than the plumbing code requires, and it is more expensive than the other devices. So, we're going to select the DCVA for backflow protection.

STEP 7

SIZE VALVES

We know that we need 6 electric control valves, and by looking at the manufacture's specifications we can determine that we want to use 3/4" electric valves. That size will be adequate to accommodate the flow rate of the zones on this system. We're going to get valves with a flow control that will allow us to fine tune the entire zone once it is installed. The least expensive 3/4" valves will have no flow control, but quality is important and we want as much control of the water as we can reasonably get.

We can install the valves as two (2) *valve manifolds* of three (3) valves each. Each manifold will fit into a 12" standard valve box. A valve manifold is an assembly of control valves. Installing valves and valve boxes will be explained in more detail in the Construction chapter.

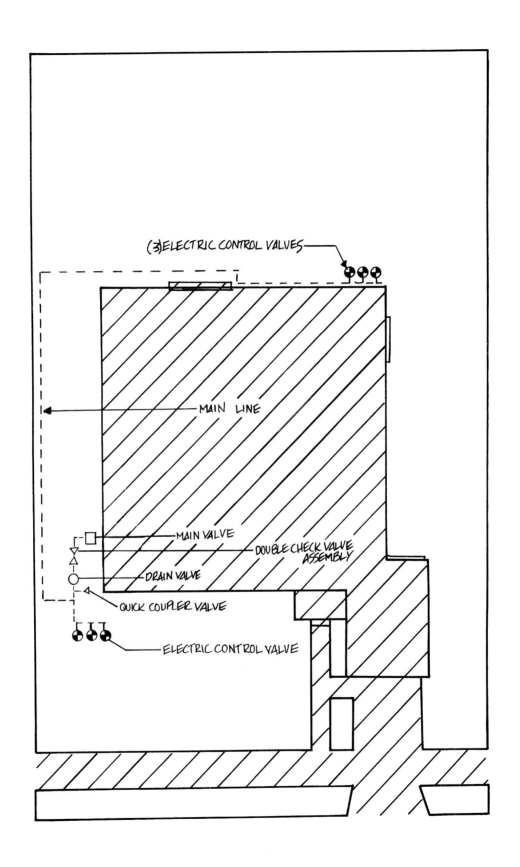

2-11. Schematic drawing of the mainline, from the point-of-connection to the control valves.

2-12. A 1" PVC mainline enters a valve pit and ends in a "valve manifold" of four electric control valves.

STEP 8
SELECT CONTROLLER

Because we know we will be using six electric control valves, we can now select a six station controller. We can put the controller inside the garage on the NE wall, where there is a convenient 117 VAC receptacle to plug into.

STEP 9
PIPE SIZING

Pipe sizing is the process of determining what is the correct diameter of pipe to use for a particular section of the sprinkler system. We do this for several reasons.

Water flowing in a pipe travels at a certain velocity to get a given volume of water through a pipe of a particular diameter in a fixed amount of time. For example, if I run a group of sprinklers with an aggregate demand for water of 15 GPM, and the water is flowing in a 3/4" pipe, the water must travel at 9.02 feet per second (FPS) to pass through the pipe at that volume. However, if I increase the diameter of the pipe to 1", then the water only needs to travel at a velocity of 5.57 FPS, in order to pass the same volume of water. The significance of this is that we want the water traveling in our pipes at close to 5 FPS. The reason for this is that water traveling at excessive velocities can create water hammer, which puts undue stress on the system components, and shortens the life of the system.

Additionally, the faster that water travels, the greater the amount of pressure lost to friction will be. Excessive pressure loss will result in the sprinklers not operating correctly.

I don't want to digress into a bunch of technical details and formulas here because I don't think that would be useful for a homeowner project. I'm just going to suggest that when you build your sprinkler system, as a rule of thumb, you use 3/4" pipe for anything up to 10 GPM and use 1" pipe for flows over 10 GPM and up to 20 GPM.

To determine the flow in a given section of piping, start at the head furthest away from the control valve and work back towards the valve, adding in the GPM of each sprinkler as you go. The piping at the far end of the line will be 3/4". When the GPM you are adding up gets to 10 GPM, the pipe size will bump up to 1".

❧ One size fits all ❧

When you buy your sprinkler parts, you will be buying fittings that are sized for the pipe you are using. Three-quarter inch pipe will require 3/4" fittings, 1" pipe will require 1" fittings, and so forth. Contractors use the smallest pipe possible because they use lots of pipe every year and using pipe that is larger than necessary is very wasteful. This is especially true on large jobs, commercial work and golf courses where large diameter pipe is very expensive.

For a homeowner building one small sprinkler system, you need to decide if it's worth it to you to size pipe and use different size fittings, or if you want to simplify the process and use just one size of pipe and fitting for everything.

Just suppose I build a sprinkler system with flows at various points in the system that range from 1 GPM to 20 GPM. I can either buy three different sizes of pipe and fittings (1/2", 3/4" and 1") and size the pipe "correctly", or I can do the whole thing in 1" pipe,

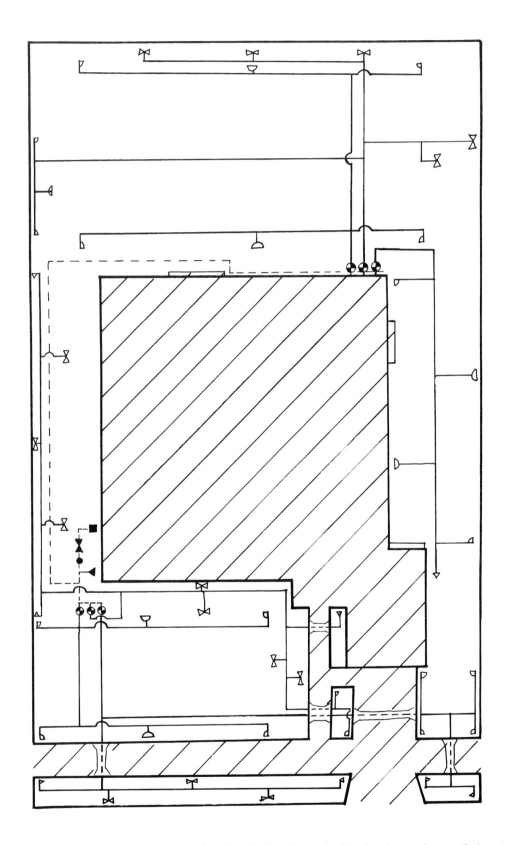

2-13. This is the complete irrigation plan. Note that the plan is schematic. The heads are close to their actual locations, but the piping is drawn to represent which heads are connected to which valve rather then the actual physical location of the pipe. When the pipes are installed, some of them will share a common trench. The pipes are shown separated in the drawing for clarity.

buy just one size of fittings, and have a lot less left over pipe and fittings when the job is done. My point is that what might be cost effective for a contractor (i.e. careful pipe sizing) may actually not be for a homeowner buying small quantities for a one time project.

Just remember that you need to accommodate your maximum flows. If you build your system with one size of pipe, you have to use the largest required size. In our example above, you could use all 1" pipe, but you could not use all 1/2" or 3/4" pipe.

At the end of this book you will find tables for pressure loss and velocity that will give you more technical detail if you want to calculate those things for yourself.

STEP 10

"WORST HEAD" ANALYSIS

All sprinkler heads have a pressure range that is specified by the manufacturer as being the correct pressure range for that particular head. When we check our sprinkler design, we do what is sometimes called the "worst head" test. We start with the static pressure at our point-of-connection, and subtract out all the pressure loss caused by the friction and elevation changes between the water meter, or other water source, and the head furthest from the water source. If we have more than the minimum amount of pressure required to operate this "worst head", then we know that all the other heads will have enough pressure to operate correctly.

If we neglect to do this calculation, then we run the risk that not all of our heads will pop-up and spray correctly when the sprinklers are operating.

One of the most common mistakes that I have observed homeowners make is to try to run too many sprinkler heads on the same valve and/or to use undersized piping. The result is that sprinklers that are supposed to spray a particular distance only spray a fraction of that distance, or don't pop-up at all.

As an example of "worst head" friction loss analysis, take a look at Figure 2-14. The head at the end of the line (circled) looks like it might be the weakest head in the system, based on the size of the zone and the distance of the head from the meter.

To do our analysis, we want to take our water pressure at the source, which in this case is the water meter, and determine how much pressure will be lost to friction by the time the moving water gets to the designated "worst head". We will do this by using the standard tables provided for you in this book.

To read the tables, we need to determine the GPM of the zone in question. By adding up the GPM of all the heads on this particular zone, we find that our total is 8 GPM. When we look up the pressure losses from our meter, valves, and pipe we will read across the 8 GPM row because that is how much water will be flowing through those components when the sprinklers are running.

Take a look at the plan, and imagine the route the water will take as it goes from the meter to arrive at the last sprinkler head.

Here is a list of each fixture and piping that the water will travel through, along with the pressure loss from friction that will occur. The pressure loss figures are taken right off the tables in the back of this book. These are standard numbers.

At 8 GPM:

1. 5/8" water meter. 2.3 PSI.

2. 35' of 1" type L copper tube. This is the water supply from the meter to where we will make our point-of-connection for the sprinklers. .75 PSI.

3. 3/4" angle valve. Our sprinkler main valve. 1.2 PSI.

4. 3/4" double check valve assembly. Our backflow preventer. 4.0 PSI.

5. 95' of 1" class 200 PVC pipe. The sprinkler mainline from the DCVA to the control valve. .97 PSI.

6. 3/4" electric control valve. .38 PSI.

7. 60' of 3/4" class 200 PVC pipe. The lateral piping from the control valve to the last head. 1.93 PSI.

8. Miscellaneous fittings. Approximately 1 PSI.

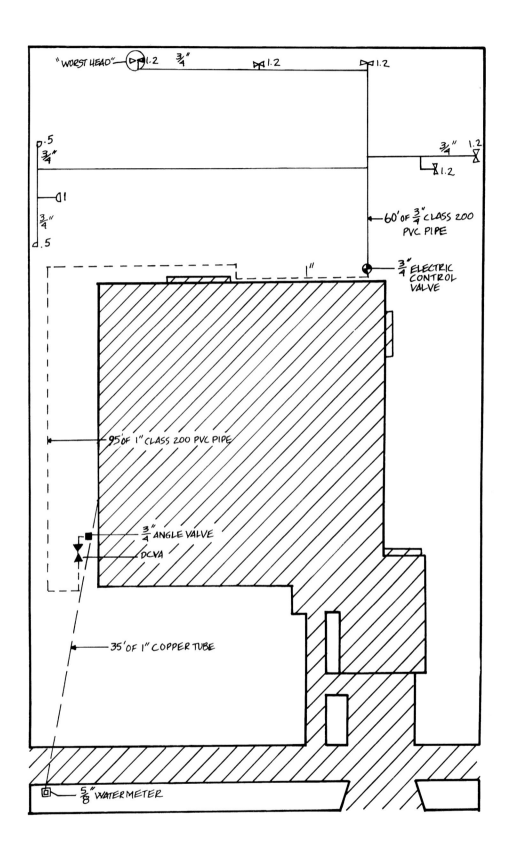

2-14. The "worst head" is circled. The water from the meter has to travel further to reach this head than to any other head in the sprinkler system. This drawing shows the route the water takes to get from the meter to this "worst head."

When we add it all up we find that we are losing about 16 PSI between the meter and the last head. That will be a conservative number because we have used 8 GPM for the sake of simplicity. In reality, the piping closer to the far end of the line is carrying less water than that.

If we take our beginning static pressure of 55 PSI and subtract our pressure loss of 16 PSI, we get a remainder of 39 PSI. That means that the working pressure at the "worst head" will be approximately 39 PSI.

You will want to consult the manufactures specifications to see what the optimum operating range is for the sprinkler heads you select. Most residential heads operate in the 25 to 75 PSI range. Our "worst head" will work great so we won't bother crunching any more numbers for the rest of the system. I'm convinced this design will work. Our working pressure and flow will be adequate at every head. If we did not have enough pressure at this head, we would have to possibly increase a pipe or valve size, run fewer heads, or otherwise re-design until we had a workable plan. If we had too much pressure, we would need to install a pressure reducer.

As it is, this plan will work. When you get a plan for your own sprinklers, simply start at the meter and work your way to the "worst head", making a list of each component and it's size as we did in the example. Use the tables to look up the pressure loss for each component, and subtract the total pressure loss from the beginning static pressure at the water source. The number you get will be the working pressure at the "worst head".

This check is the last step in the design. It only takes a few minutes and is much easier than having to make corrections to a sprinkler system that is already installed.

Once you have a workable plan, you can start building your sprinkler system.

❧ Design Worksheet ❧

Water meter location:_____

Water meter size:_____

Water meter capacity (GPM)_____

Maximum safe flow for irrigation (75% of meter capacity)_____

Static water pressure (In PSI):_____

Flow (GPM) from existing outlet (hose bib, etc.)_____

Service line to house: Material (plastic, copper, steel, etc.)_____

Size of service line?: (diameter, 3/4", 1" etc.)_____

Maximum flow through service line (In GPM):_____

Condition of service line? Suitable for cross-connection?_____

Possible location of point-of connection. Where could you potentially tap into your water system to supply the irrigation? (At meter, at shut-off valve by foundation, in basement, etc.)
_____

Final landscape plan? Yes____No____

Select and layout heads/nozzles_____

Count total irrigation water demand by adding up total gallonage of all heads. Total GPM lawn areas_____ plus Total GPM shrub areas_____ equals TOTAL GPM_____ . Then divide total GPM by available water_____= Minimum number of zones needed____

Other factors affecting the total number of zones needed?_____

Optimum number of zones_____

"Worst head"

List pressure loss to your worst head.

GPM of worst head zone_____ (Add up GPM of each nozzle on this zone).

Beginning static pressure_____

Find pressure loss from tables:

1. Meter_____

2. Water line_____

3. Main valve_____

4. Backflow_____

5. Control valve_____

6. Mainline pipe_____

7. Lateral pipe_____

8. Elevation rise in vertical feet_____ x .433 = _____ PSI loss.

9. Elevation drop in vertical feet_____ x .433 =_____ PSI gain.

10. Difference between line 8 and line 9_____(net pressure loss or gain from elevation)

11. Total lines 1 through 7_____

12. Add loss or subtract gain from line 10_____

13. Total pressure loss to worst head_____

14. Beginning static pressure_____

15. Minus total pressure loss (from line 13)_____=

16. Working pressure at worst head_____

Recommended working pressure for that head_____

Is plan workable? Yes___No___

If no, re-design sprinklers. If yes, proceed to next chapter.

Chapter 3

Low-Volume Irrigation

In Low-Volume Irrigation, *you will learn the water-saving principles of drip and micro irrigation. You will learn the components of low-volume irrigation that are different from the sprinkler hardware used in conventional overhead spray irrigation. You will also learn how to design low-volume irrigation into your sprinkler system.*

Low-Volume Irrigation

Low-Volume irrigation is a method of watering individual plants, or groups of plants, by applying water at low pressure and low volume to a very specific target area. By applying water only to the necessary root areas of the plants at close range, water is used very efficiently with only minimal amounts of water being lost to evaporation, wind drift, overspray, and run-off.

Because the delivery of water is very controlled, water is not wasted irrigating the spaces between the plants.

Many people refer to low-volume irrigation as drip irrigation. Drip is certainly one kind of low-volume irrigation, and there are also other watering systems in addition to drip, such as micro-spray, which would fall under the broader heading of low-volume systems.

Let's take a look at the major components of a low-volume system.

Pressure reducer

The pressure reducer is a device installed in-line for the purpose of regulating the water pressure of the low-volume system. Most low-volume components operate in a pressure range that is less than 30 PSI, although there are some (low-volume components) that operate at pressures as high as 50 PSI. That means that if your water service provides adequate pressure for a conventional sprinkler system, it is probably above the optimum pressure we want for low-volume irrigation. This also means that if you do have low pressure at your site already, low-volume irrigation may provide a solution for you when there is not enough pressure to run conventional sprinkler heads.

A pressure regulator will convert a range of inlet pressures to one constant outlet pressure. Some pressure regulators are adjustable as to your selection of an outlet pressure, and some pressure regulators only provide one preset outlet pressure.

A large pressure regulator, like the kind that might be on your water service line if you live in an area where there is high water pressure, will typically have an adjustment to regulate the discharge pressure. For example, if the incoming water pressure is 110 PSI, you can set the outlet for 50 or 70 PSI, or whatever you want for your domestic water use.

The smaller pressure regulators, like those that will typically be found on a drip zone, give one fixed outlet pressure of 30 PSI, or whatever the manufacture has determined is the optimum pressure for that line of drip components.

Filter

We install a 100 to 200 mesh filter at the inlet to a low-volume system to keep small particles that may be in the water from entering the downstream components. We do this because the small diameter tubing and emitters used in low-volume irrigation can easily become clogged. This happens because a drip emitter does not flow with enough force to clean itself very well. Having a fine mesh filter at the inlet to the system reduces clogging, which reduces the amount of overall system maintenance required.

If you are installing low-volume irrigation, there are two possibilities. Either you have a system that consists only of low-volume irrigation or you have a system that consists of a mix of conventional spray zones and low-volume zones.

A low-volume system will typically require a pressure reducer and a filter in addition to the standard components of a conventional system. If the low-volume system is one or two zones of a sprinkler system that also has conventional spray zones, then the filters and pressure reducers would be installed with the low-volume control valves. If the entire system is a low-volume system, then it would make more sense to use one pressure reducer and filter at the inlet to the entire system.

Valves

The manual valves in a low-volume system will be the same as in a conventional sprinkler system. The electric control valves in a low-volume system need to be of a type that will shut off at low flows. Because of their construction, not all electric valves operate correctly at very low GPM's. Make sure you purchase a valve that is compatible with a low-volume application.

Pipe

Hardpipe

A low volume system may be installed with the same PVC pipe that is used in a conventional sprinkler system. When PVC pipe is used in a low-volume system it is often referred to as *hardpipe*.

Softpipe

There are a variety of flexible tube pipes manufactured for low-volume sprinklers. This flexible tubing is often called *softpipe* to distinguish it from the ridged PVC pipe. Softpipe for low-volume systems will be manufactured from polyethylene (PE) or vinyl. Some of these softpipes are solid, and convey water to a specific emission point. Some of these softpipes are slotted, and disperse water along their entire length, like a soaker hose.

Softpipe for a residential application will come in 1/2" and 1/4" diameters. The 1/2" diameter softpipe, also called *drip tube*, will be used as lateral piping. The 1/4" tube, also called *distribution tube* or *spaghetti tubing*, is used to carry water from the lateral piping to each plant.

In a typical installation, a 1/2" drip tube will extend the length of a planting bed. The 1/4" distribution tube will then run from the drip tube to each individual plant. The 1/4" tube is connected to the 1/2" tube by using barbed connectors which are pushed through a hole made in the wall of the drip tube.

There are tools made for the purpose of punching a small hole in the drip tube that is the right size for inserting a barbed connector or barbed emitter. A small drill bit can be used for the same purpose.

When PVC pipe is used instead of 1/2" softpipe for lateral piping, then the 1/4"

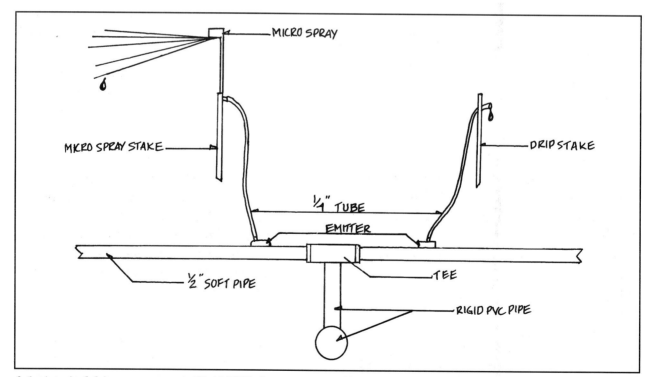

3-1. A typical drip arrangement. Rigid PVC pipe connects to 1/2" PE pipe. The 1/2" pipe extends along the length of the bed. Emitters are then punched into the 1/2" pipe. The 1/4" spaghetti tube connects to these emitters and goes to the individual plants. The 1/4" tube then drips at the root zone of the plants. The 1/4" tube could also end in a micro-spray nozzle to water a small area of flowers or groundcover.

distribution tubing is connected to the PVC pipe by using multi-outlet emission devices which connect to the PVC pipe by a threaded fitting.

EMISSION DEVICES

Emitters regulate the flow of water through the 1/4" tube. An emitter may or may not be pressure compensating. A pressure compensating emitter is designed so that all the emitters on a length of pipe emit a uniform amount of water, regardless of changes in the working pressure along the run of pipe caused by friction loss and elevation change. An emitter that is not pressure compensating serves as a barbed connector between the 1/2" tube and the 1/4" tube, but does not regulate the flow as precisely as does a pressure compensating emitter.

Some emitters have multiple ports. This type of emitter is typically installed on PVC pipe and connects the PVC pipe to 1/4" distribution tubing. Although used less frequently, there are also multiple port emitters that connect to 1/2" drip tube with a barbed connector.

Emitters are sized according to how much water they discharge in a given amount of time, expressed as gallons-per-hour (GPH). Our conventional sprinklers, you recall, are sized in gallons-per-minute (GPM). One GPM equals 60 GPH. A sprinkler system with a capacity of 15 GPM is equal to a low-volume capacity of 900 GPH (15 x 60 = 900).

Emitters will discharge water in amounts ranging from .5 GPH to 24 GPH. Emitter sizing is covered under the *design* heading in this chapter.

Micro-sprays are small nozzles (analogous to the nozzle on a conventional spray head) which perform at very low volume and pressure. Micro-sprays are usually mounted on plastic stakes. They are appropriate for areas of ground cover, flowers and other dense plantings, where the use of individual distribution tubes to each plant would not be practical.

Soaker hoses for low-volume systems are versions of the soaker hose you are already familiar with. Some of the low-volume hoses are designed so that the entire hose fills up with water, and then emits the water evenly along its entire length. In my opinion, this is a good feature because at very low flows there could be a tendency for a conventional soaker hose to emit all the water at the end closest to the valve with only a small amount, if any, getting to the plants at the far end of the soaker. Even with soaker hoses that are supposed to emit water evenly, you should be careful when using these products on a slope. Water distribution tends to be uneven with these products when there is any significant elevation change along the length of the run.

FITTINGS

The 1/2" drip tube is joined by the use of either *compression* or *barbed fittings*. With a compression fitting, the tube is pushed into the fitting. Compression fittings are designed so that once the tube is inserted, it cannot be pulled back out. With a barbed fitting, the tube fits over the outside of ridges on the fitting. The ridges keep the tube from slipping off. Some 1/2" barbed fittings require the use of a hose clamp, and some do not. The 1/4" drip tube is joined by the use of barbed fittings that push into the tube and do not require a clamp.

There are adapters available to go from a threaded hardpipe fitting to a compression or barbed softpipe fitting. These are called transition fittings.

BACKFLOW

A low-volume system has the same backflow prevention requirements as a conventional sprinkler system When selecting and installing backflow prevention for your low-volume sprinklers, follow the same guidelines we used when discussing conventional spray sprinklers.

HOSE BIB CONNECTION

Because of the low pressure and flow requirements, you may be able to connect your low-volume system to an outside hose bib as a water source instead of connecting to your water line. If this works for you, remember that you still need to provide a backflow preventor. Also, if you are in a climate where there is a possibility of freezing temperatures, you will want to provide drainage for any piping that is above grade and connected to the hose bib.

❦ Low-Volume Design ❦

Let's refer back to the landscape plan (Fig. 2-4) that we used for our design of a conventional sprinkler system. Just suppose we design a low-volume zone to water the shrubs in that same plan instead of the conventional spray irrigation that we put in our original design. Then we can compare the two methods. We can use drip emitters for all the plants in the areas we previously designated as shrub/sun exposure except for the small bed between the garage and the entry walk. Because that bed will contain annual flowers, we would irrigate it with micro-sprays. The reason for that is because in a densely planted bed, we don't want a tangle of 1/4" tubing going to every plant. The micro-spray nozzles will cover the area without the tube clutter. In the bed area that we previously designated as shrub/shade exposure, we can leave on a separate valve and irrigate with spray heads as before. This is because that area contains a mass of groundcover between all the shrubs and so we want to water the entire area. We could also water this area with micro-sprays, but the area is so large and densely planted that it is just as well to use the spray heads.

DETERMINE SOIL TYPE AND DISPERSAL PATTERN

One of the first things I would like to know is how many emitters I will have to place around each shrub and tree. To answer that question accurately, we will need to take a look at the soil we are planting in. The reason for that is because the soil type will determine the pattern of water dispersal from each individual emitter, and that will determine the distance apart that the emitters can be placed,

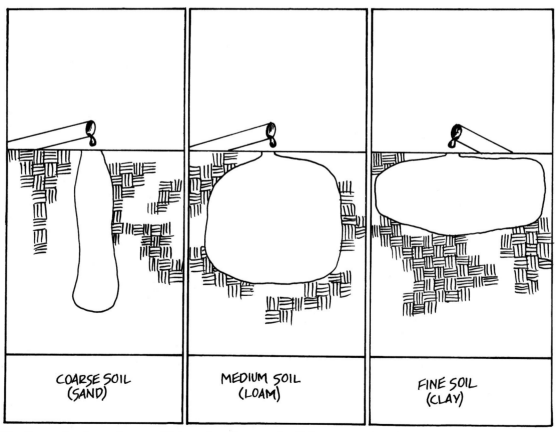

3-2. The dispersal pattern of water will depend on the type of soil. In a course (sandy) soil, the water moves down in a narrow pattern, with little lateral movement. In fine (clay) soil, the water spreads widely, and has only shallow penetration.

which will determine how many emitters we need to do the job.

Figure 3-2 shows the different dispersal patterns in three broad categories of soil. For a more detailed description of the soil types, refer to Chapter 5 in this book.

You can see that in clay soil, we get a wide and spreading pattern because the water runs off, instead of penetrating into the ground. Clay soil will require the fewest number of emitters. That's not necessarily good. Dense soils are not the best for growing things. It just means that, because of the dispersal pattern of water in this type of soil, the emitters can be spaced further apart. Regardless of the soil type, we want an emitter spacing that will wet a minimum of 70% of the surface over the root area. With a clay soil, you will select an emitter with a slow discharge rate. The idea will be to apply water very slowly for a long period of time, so that the water will have a chance to soak into the hard, dense soil.

Sandy soil has a very different dispersal pattern. In sandy soil, the water tends to percolate down directly under the emitter and not disperse laterally. You can see that in sandy soil we will need our emitters closer together and therefore require a greater number of emitters to do the job. In sandy soil, like clay soil, we will choose emitters with a low discharge rate. In sandy soil, the water will not run off, but if it is applied too fast it will percolate into the soil deeper than the roots can use it. We want to keep all of our water at a depth where it is usable.

In a loamy soil, we have a dispersal pattern that is wider than sand and more narrow than clay. An emitter in a loamy soil will wet a spot that is 2' to 4' in diameter. After that, additional run time does not increase the area of the wetted surface as the additional water moves down into the deeper soil. If you're landscaping your yard, I hope you have taken the step of amending whatever native soil was there to begin with so that now you are planting in a loamy soil.

LAY OUT EMITTERS

Let's assume that the yard we are looking at has loamy soil. I will use 2 emitters for each small shrub and each arborvitae, and 3 emitters for each medium shrub. The boxwood hedges in front, because the individual plants are so close together, will "share" emitters, per plan. In the annual bed between the garage and the entry walk I will use the same emitters as I am using for the other plants. The difference is that the emitters in the annual bed will connect to 1/4" tube which connects to microsprays to water the entire area. The other emitters will connect to 1/4" tube which is then placed to discharge water directly at the root zone of an individual plant.

When we install this, we will position the emitters so that they discharge at a point that is half-way between the crown of the plant and the drip line.

SIZE EMITTERS/ZONE SYSTEM

My next step is to count how many emitters I have laid out. In our example, we have a total of 113 emitters. We have already determined that these are all shrubs with a similar water requirement, a similar exposure, and they are planted in a similar soil. Because of that, we will water them on the same valve if our water supply is adequate for that.

When we designed our conventional sprinklers, we determined that the water available to us at this site was 15 GPM. You may refer back to that chapter to get up to speed on that if you missed it.

$$GPM \times 60 = GPH$$

Because low-volume systems are sized in gallons-per-hour (GPH), we will use the formula given here of convert GPM to GPH.

Multiplying 15 GPM by 60 gives us 900 GPH. What that means is that we may use a number and size of emitters on a zone whose aggregate demand for water does not exceed 900 GPH.

Because it is desirable, in this case, to run all the shrubs on one zone, I will divide the available water by the number of emitters needed. Doing this will tell me what size of emitter to use if I am to water all these plants on one valve.

900GPH / 113 emitters = 7.96 GPH per emitter

By doing this calculation, I now know that if I purchase and install emitters that discharge approximately 8 GPH, I can water all the shrubs on one zone.

Run time

It may well be that 8 GPH is the right size emitter to use, but let's think about other factors that could influence our selection of emitters.

The slower the discharge rate, the longer the emitters need to run to apply the required amount of water. You are going to find that emitters are available that will discharge anywhere from .5 GPH to 24 GPH. This is quite a spread. At 24 GPH, you have a precipitation rate that is comparable to a conventional sprinkler system and you may be watering for a few minutes. At .5 GPH, your are watering at a very slow rate (that's why they call it drip) and, depending on the soil and exposure, you may be figuring your run times in hours, not minutes.

When you size your emitters by dividing the available water by the number of emitters, the next thing to do is to determine how long the zone will need to run to water the plants at that particular rate. You can experiment by plugging a few emitters in then checking the soil to see how the depth and spread of the water pattern is and how wet the soil is at given run times.

Whatever the discharge rate of the emitters is, you can probably make them work by adjusting your run times. Since there are so many variables from site to site, you will want to monitor your new installation closely and make adjustments as needed until you get it right for your plants. You may find that you want to fine tune the system by changing some of the emitters to higher or lower discharge rates, to make individual plants more comfortable.

Once you get familiar with your system, you will develop a feel for the relationship between the rate at which the emitters discharge water and your run times. As your plants grow and change, your sprinkler system will need to grow and change occasionally, also.

Layout piping

Referring back to our plan now, we will select one of the valves in the front yard to become our drip valve. Located with this valve will be a pressure reducer and filter. Since the rest of the sprinkler system will remain a conventional sprinkler system, we will install the filter and pressure reducer on this one zone only.

As the plan shows, we can come out of that valve with PVC pipe, and run PVC pipe to each group of shrubs. At 5 separate points along the PVC pipe we will "tee" off into a run of 1/2" drip tube that runs the length of each planting bed. We will keep each tee at approximately the mid-point of each run of drip tube to equalize the pressure. We want to keep our runs of 1/2" drip tube to under 100' to avoid excessive pressure differences caused by friction loss from one end of the tube to the other.

Once I have the 1/2" drip tube going the length of each planting area, I can plug in the individual emitters to which I will then attach the 1/4" distribution tubing which will go to the individual plants.

❧ Installation ❧

Just suppose we were to install this drip zone to water the shrubs when we installed the sprinkler system in the construction chapter.

If that were the case, we would proceed in exactly the same way except that we would not install the lateral drip tube until after everything else had been installed and backfilled and the plants installed. That is because we are going to lay the drip tube and distribution tube on top of the ground and we don't need it underfoot when we're doing the landscaping.

After the plants are installed, we will run

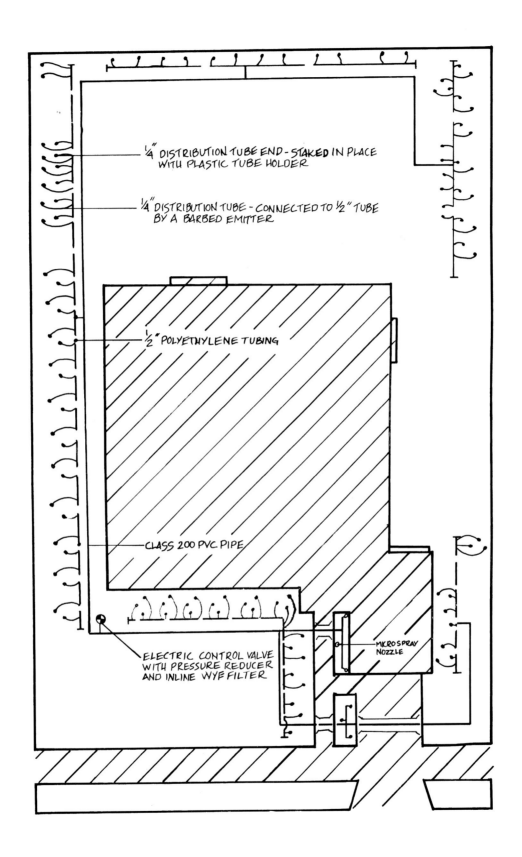

3-3. The layout for watering the shrubs in our plan with drip and micro-spray irrigation.

the drip tube close to the shrubs. Then we will punch holes in it with a tool designed for that purpose and push in some pressure compensating emitters that will give us 8 GPH apiece. We will pick the emitters based on the manufactures specifications and our design requirement that we arrived at earlier.

Once the emitters are plugged into the drip tube, we will push a length of 1/4" distribution tubing over the barbed outlet of each emitter and run the required number of tubes to each plant. We will position the discharge end of the tubes so that they are equally spaced around each plant, and also locate them so that they are half-way between the drip line and the crown of each plant.

You could leave the distribution tubing on the ground. But, I prefer to take one more step and stake the end of each tube in position with a plastic stake designed to hold the tube in place with the discharge end a few inches above the ground.

The reason why I like to add stakes is so that the tube stays is position, and also so that the discharge from the tubes is easier to inspect visually. I want to check the installation occasionally with the water on to confirm that everything is working. If the tubes are buried where they discharge water, I won't know if they're working or not until I see a plant stressing.

Once that is done, we are also going to pin the 1/2" and 1/4" tubing to the ground with wire stakes made for this purpose, wherever we think it may need to be secured.

The micro-sprays will be installed in much the same way. The 1/4" distribution tube to a micro-spray will connect to a plastic stake which supports the micro-spray nozzle.

After our drip zone is in place, we will cover the tubes with a few inches of bark or other mulch to hide the tubes and conserve water.

My guess is that this installation would save more than half the water that would be used to water the shrubs with the spray heads. Also, it will be much easier to control the weeds because the ground between the plants won't get watered.

There are a lot of low-volume gizmos and gadgets available and some work better than others. A trip to the sprinkler store or garden center will give you more ideas for your own low-volume system. This is a good way to water container plants and hanging baskets on the patio, and will also work in the vegetable garden.

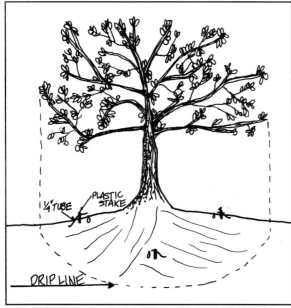

3-4. Stake the ends of the 1/4" spaghetti tube halfway between the crown of the plant and the drip line of the plant.

3-5. Spaghetti tube staked at the base of a rose bush. This spaghetti tube ends in a "bug cap" which will discourage insects from entering the end of the tube and clogging it, while allowing the water to be dispersed.

Chapter 4

Construction

In Construction we will get down to the practical nuts and bolts of building your sprinkler system. I will give you a pre-construction checklist to help guide you through the process of obtaining a permit, getting the underground utilities located, assembling a bill of materials and buying parts. We will decide what tools and equipment are needed, and establish a "critical path" to keep the project on track. We will proceed, step-by-step, through the entire installation process- layout, trenching, installing pipe, valves, heads, wire, valve boxes and controller - including many "hands-on" tips and shortcuts to achieving a professional quality installation.

❦ Pre-construction: ❦

We are just about ready to begin installing our sprinkler system. there are several things to do before we break ground.

1. Call for utility locates. *This is very important.* Buried underground utilities are a serious hazard and ***you absolutely must have them located before you do any digging.*** Some areas have a one call service that will notify all the various utilities (electric, gas, water, phone, cable) or you may need to call them yourself. However it happens, *it's up to you* to get the utilities located before you dig. This service is available free everywhere that I know of. However, if you hit a underground utility because you neglected to have it located, *it can cost you a lot,* not to mention the potential for ***serious injury and property damage.***

When underground utilities are marked with paint or flags, the markings may not be the exact location of the buried utility. Consider that the actual wire or pipe may be several feet to either side of the locate mark.

If you are ever using a trenching machine in an area where there may be underground hazards, my suggestion is to very carefully dig the area in question by hand. Once you have the hazards exposed and know exactly where they are, you can machine trench the rest of your site without hitting them.

The drain lines that go from your roof gutters to the street are usually very shallow, but won't get located by the utilities. Even your own water line from the meter to your house probably won't get located, because a utility doesn't own those things, you do. You might be able to guess the route your drain lines take and be able to avoid hitting them by hand digging the area and exposing enough pipe to determine the path of the pipe. But, because they are usually shallow, it's not unheard of to hit one with a trencher. If you do hit a drainline, dig out the damaged section and repair it. Chances are that your drainlines will be ABS plastic pipe which can be cut and glued just like PVC pipe. Use a glue made specifically for ABS plastic if you find yourself in this situation. There is also a rubber coupler manufactured for drain pipe. The rubber couplers clamp on to the sections of pipe with metal hose clamps.

2. Get the plumbing permit you will need for your backflow device. Call your local building codes agency for specific information on this. The process usually consists of filling out a form and (big surprise) writing a check. You will get a permit number which you will use after the backflow device is installed, when you call for an inspection. Different states, different cities, different water districts will all have their own water quality, plumbing and possibly other regulations that may affect how you proceed with your project. Inquire locally.

3. Let's make a bill of materials so that we can have what we need on hand when we're working, and also so that we can know what to expect for costs. A bill of materials for the plan in figure 2-13 might look something like this:

Item	Size	Qty.	Price Ea.	Total
Brass angle valve	3/4"	2		
Double check valve assembly	3/4"	1		
Quick coupler valve	3/4"	1		
Electric control valve	3/4"	6		
Plastic valve box	12"	2		
Plastic valve box	jumbo	1		
Plastic valve box 6" extension	jumbo	1		
Electric controller	6 station	1		
Low-voltage wire	18-7	170'		
Pop-up spray heads	4"	23		
Pop-up spray heads	6"	10		
Rotor heads	4"	12		
Class 200 PVC pipe	3/4"	540'		
Class 200 PVC pipe	1"	140'		
Schedule 40 PVC pipe	1"	10'		
Schedule 40 PVC pipe	3/4"	30'		
Class 200 PVC pipe	2"	4'		
Class 200 PVC pipe	1-1/4"	30'		
PE pipe	1/2"	100'		
Brass hinged cap	2"	2		
Sub-total .... $				
Miscellaneous (25%)				

Total Material $

Your bill of materials may not even have prices at first. Unless you already have price lists available, just fill in the item, size and quantity. You can fill in the prices later. In the miscellaneous category you will cover fittings (Ells, tees and so forth), wire nuts, waterproof splices, glue, rags, saw blades, and so on. Buy those things as needed. After I get prices for the major parts, I will add 25% to the total cost. That 25% will cover the miscellaneous parts and supplies.

I'm not a gambler, but if I was I would bet that your first sprinkler system will cost more and take longer to build than you think it will. Of course, we could also say that about every other home improvement project you will ever do, so don't be discouraged, just be prepared.

OUR PROJECT

The sprinkler system you and I are going to install together (on paper anyway) is Figure 2-13. We designed this sprinkler system in the chapter on design. This is a new home, the grades have been established and nothing is planted yet. We're going to irrigate for the proposed landscape which is Figure 2-4. Assuming all the pre-construction steps have been taken, let's get started.

CRITICAL PATH

Before we plunge into the project, it will be useful to rough in a schedule of how we anticipate getting the job done. By organizing the tasks in the right sequence we can make our job easier – saving time, money and wear and tear on ourselves.

For example, to install the sprinkler system in our plan, our game plan might look something like this:

Day 1: Lay out head and valve locations; dig out point of connection; install main valve and double check valve assembly; call for plumbing inspection (if day 1 falls on a weekend, call for inspection on the following Monday).

Day 2: Rent trencher; Do all machine trenching and return trencher; Install wire and pipe; Backfill.

Day 3: Flush mainline; install valves; plumb connections at valve pits; install valve boxes;

Day 4: Install controller; wire valves and controller; flush lateral lines; set heads; adjust nozzles; clean-up.

Will this be the scenario for every job? Of course not. There are any number of variations, depending on your individual situation. The point is, think the job through in advance and put together a game plan. You don't want to pay extra rent on a trencher because you make the mistake of picking it up two days before you are actually ready to use it.

MATERIAL AND SUPPLIES

Buy the materials now, so that you will have them on hand as you need them. If you live in the cheapest home you could buy, drive the cheapest car you could find, wear the cheapest clothes, and eat the cheapest food, then by all means buy the cheapest sprinkler parts you can find. However, if value and quality and maybe getting something that actually works are important to you, then look around a little bit. Shop and compare value, not just price. Is the $40.00 controller just as good as the $125.00 controller? Really? Talk to the sales people. Compare warranty and return policies. Can you expect any advice or technical support after the sale? How are we going to get the pipe to your house when it's stocked in 20' lengths?

Let's also get our tools and supplies lined up. Keep in mind, this is what we need to do this project. Your own could well be different. The point is, think this job through with me and then you can transfer the process to your own situation.

Here's what I'm thinking, just off the top of my head, is what we'll need for supplies:

1. PVC glue. To glue our PVC pipe and fittings. We want a medium-bodied, fast-drying glue.

2. PVC primer. To use when making glued connections on PVC pipe.

3. Teflon tape. To wrap our threaded fittings.

4. Pipe dope. For the threaded fittings that

will make swing joints for our quick coupler. We want a product that is compatible with PVC pipe.

5. Caulking. To seal the hole in the garage wall after we install the low-voltage wire.

6. Rags. To clean and dry the pipe and fittings before gluing, and to wipe off excess glue.

7. Marking paint in a aerosol can. For marking bed lines. The same kind of paint the utilities use for putting paint marks on the ground. We'll pick a different color than the utilities used, to avoid confusion.

8. Sprinkler flags. For marking head locations.

Since we have to make our point of connection in a copper waterline (for this particular project, your own may be different) we will also need:

9. Lead-free solder, approved for potable water. For making the point-of-connection in the copper waterline.

10. Soldering flux.

11. Emery cloth/sandpaper

And, for tools, let's start with:

1. Round-point shovel. For digging large holes.

2. Trenching shovel. The kind that looks like a clam digger. the narrow blade is good for digging trenches.

3. Flat point shovel. For blading dirt off concrete surfaces, backfilling and grading.

4. Pick. You might need it if the ground is hard or rocky. (Wear eye protection).

5. Sledge hammer. For driving pipe under sidewalks. (Wear eye protection).

6. Hacksaw. A saw with a 6" blade is a good size. For cutting PVC pipe.

7. Tubing cutter. For cutting copper tube.

8. Propane torch. For copper pipe. You'll need the regulator and a small tank of propane or mapp gas. If the regulator is not self-igniting, you'll also need the striker.

9. Power drill. For making a hole in the garage wall for our wires.

10. Wire cutters/strippers. For splicing wires.

11. Large slip joint pliers.

12. Pipe wrench.

13. Screwdrivers.

14. 25' tape measure.

And, for your safety, I recommend eye protection (safety glasses), ear plugs (when operating equipment), leather footwear, and gloves.

Layout

The first thing we're going to do is to use our marking paint to paint the bed lines on the ground. To do this, we're going to take our landscape plan and find some fixed reference points, such as the corner of a building, and scale off the distance to a point on the landscape plan. Then, using our tape measure we'll transfer that to the actual location on the site and mark it with a dot of spray paint. When we have enough of these points located, we can use the spray paint to "connect the dots". Once we are done, we will have the lawn and shrub areas clearly defined, which will make it easy to flag our heads in exactly the right place. If you don't like the look of your bed lines the first time you paint them out, it's very easy to erase the paint by scuffing your shoe over it. Then, you can redraw the lines until you have what suits you.

Now that we have our lawn and shrub areas marked, we can take our sprinkler flags and mark the head locations. I like to use a different colored flag for each zone because then I have a very good visual aid to help me decide where to place the valves and how to pipe it. Next, using our spray paint again, mark the actual locations of the valves. Now we can see exactly what our sprinkler system will look like. If the bedlines changed at all from the plan, we can layout the heads accordingly by moving flags.

The design work we did earlier on paper can also be done on the site itself. Use spray paint to mark the bedlines, use stakes or paint circles to mark the proposed plant locations, and use the sprinkler flags to mark head locations. This might even be a better way for you to do your design if you have trouble visualizing the site from a drawn plan.

Trenching

Call ahead and reserve a trencher at your local rental yard. When you pick the trencher up, don't hesitate to ask for operating instructions. All the trenching on your project will be either machine trenching or hand trenching. Plan the job so that you can do all of your machine trenching at one time and return the trencher as soon as possible. Of course, you need to know if you are being charged hourly, or by the day. If having the machine for five hours is going to cost the same as having the machine for twenty-four hours, then obviously there's no reason to hurry back with it.

We want to do a minimum amount of trenching. To facilitate this, we can put more than one pipe in a single trench.

We are going to have our mainline at a depth of 12-18" below grade and our lateral lines will be 9-12" below grade. To accommodate multiple pipes in the same trench, we will start with slightly deeper trenches so that the top pipe is still at or below our minimum depth. Figure 4-2 shows how to layer multiple pipes in a trench. When putting more than one pipe in a trench, separate the pipes with a few inches of clean backfill between each pipe. If field wires are going in the same trench, we want to put the wires underneath the mainline, which will always be the bottom pipe.

In addition to trenches for the pipe, we also need to dig "valve pits" where we are going to locate our control valves, our double check valve assembly and our point-of-connection. The valve pits should be larger than the dimensions of the valve box that is being used. For our point of connection, we are going to have to dig a hole big enough to work in with a propane torch and with enough room to get the handle of a copper tubing cutter all the way around the waterline. Don't make the mistake of digging a hole that is too small to work in. Lots of bad plumbing has resulted from someone trying to work in a trench, or valve pit, that was just not big enough.

4-1. A small trenching machine, available at many rental yards can make digging easier and faster.

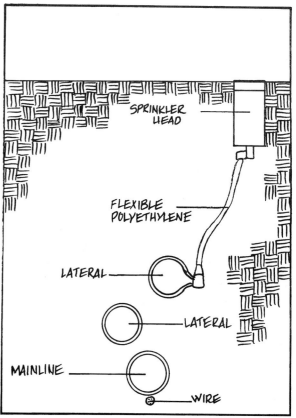

4-2. Instead of digging three separate trenches, here are three pipes laid in the same trench. The multi-strand wire is on the bottom, underneath the mainline. Two lateral lines are above the mainline, separated by an inch or two of dirt with no rocks in it.

4-3. A hole has been excavated, exposing the copper waterline that goes from the meter to the house. This is where the waterline will be cut to make the point-of-connection for the sprinkler system. The waterline is two feet below grade and the hole is dug another foot below the waterline. There is room in this excavation to work in, and to get the handle of the tube cutter all the way around the pipe. Water that drains from the pipe when it is cut, has somewhere to go without submerging the pipe.

BORING

There are several places on our plan that call for getting a pipe under an existing sidewalk. We have several options for doing this. Since the sidewalks we want to get under in this project are only 4' wide, we may be able to get by with a low-tech approach and simply dig under the walks at an angle from each side as far as we can go. Then, we'll take a 5 or 6' length of 1" schedule 40 PVC pipe and put a cap on one end. Putting the open end of the pipe as far under the sidewalk as it will go, hammer the capped end with a sledge hammer until the pipe emerges on the other side of the walk. Then, remove the 1" pipe, which will be packed with dirt, and slide a new 3/4" pipe through the hole. Cap the ends of the new pipe to keep the dirt out before shoving it under the sidewalk. If the ground under the sidewalk is particularly hard, you might have to use a length of steel pipe to pound under it the first time because the plastic will break after a certain amount of hammering.

Getting under an existing sidewalk can sometimes be very difficult, depending on the soil conditions and the amount of room one has to work in. If the muscle approach does not work for your situation, there are boring attachments that can be rented that are powered by some models of trencher. These consist of drilling bits which attach to a steel rod which is powered by a PTO arrangement on the trenching machine. The machine then drills under the walk or driveway. Check at the rental yards for what is available to you.

There are boring attachments available for use with a heavy duty electric drill that may work for you.

Sometimes a high-velocity nozzle, threaded onto the end of a rigid nipple, and attached to a garden hose will help bore under a sidewalk by using a high-velocity jet of water to blast the dirt out of the way.

If you are building a new home, you will want to make sure that the builder provides sleeves under the driveway and under the walks. The time to get sleeves in is before those areas are rocked, prior to the concrete pour.

In our project, the pipe crosses the driveway. If the driveway were not already sleeved, we would need to use a boring machine to bore under it. The alternative would be to run the pipe all the way around the house.

In extreme cases you can cut and patch the concrete or asphalt. There's always a way to do it, but some ways will cost more than others.

POINT OF CONNECTION / MAIN VALVE
COPPER PIPE

Our project has a copper waterline so I am going to assemble everything I need to sweat copper pipe. When I dig out my point of connection I want to take a close look at the exposed copper line. If the line is dented up, or on a bend, I will keep digging until I have at least a 6-8" section of clean, straight pipe to work with, and more would be even better. If the cross-section of your copper pipe is out of round, you will not be able to sweat in the tee. Therefore, it is important to find a section of pipe that is not bent or dented to work on.

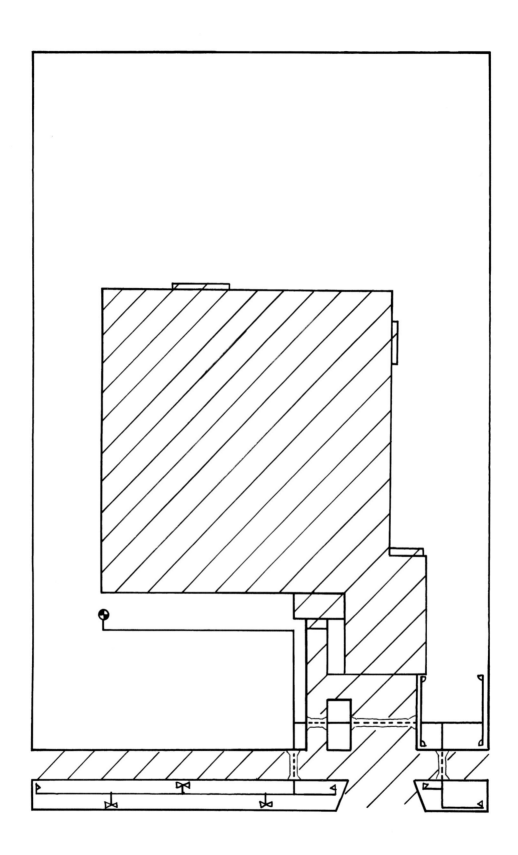

4-4. This drawing shows one possible way to pipe a zone. A zone consists of all the sprinkler heads connected to a single control valve. In this case, the driveway is "sleeved" so a pipe can be pushed through to connect the heads that are across the driveway from the valve.

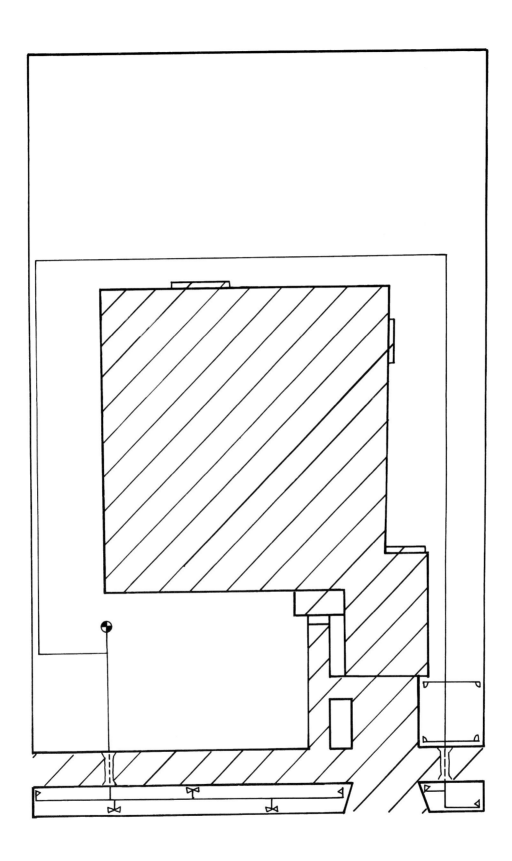

4-5. This drawing shows the same zone as Figure 4-4. The head layout is identical. In this example, because the driveway was not sleeved, the pipe goes around the house.

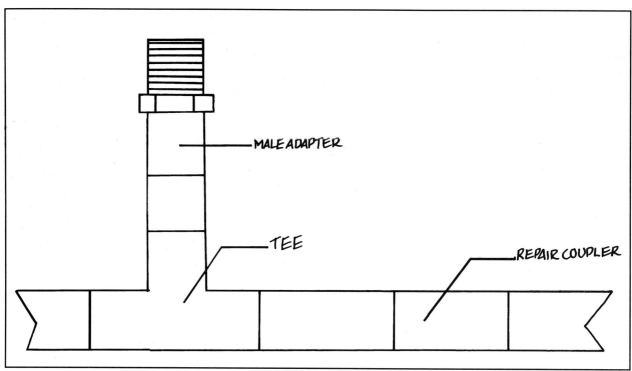

4-6. *A typical point-of-connection on a copper waterline consists of a tee and a male adapter. The tee is installed in-line by using a repair coupler.*

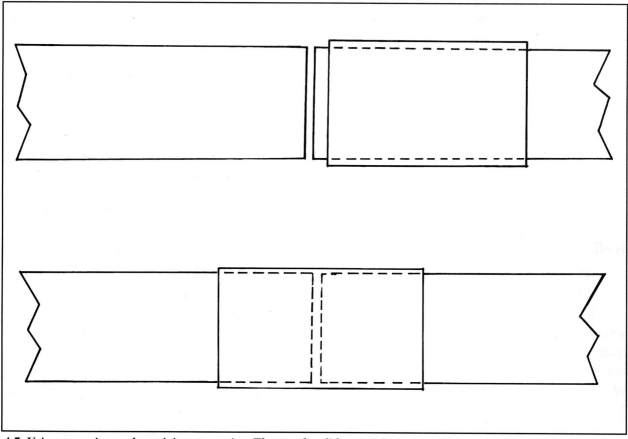

4-7. *Using a repair coupler to join copper pipe. The coupler slides completely past the end of one pipe, so that the pipe to be joined can be positioned. When the two pipes are end to end, the coupler is slid half way over each end. The solder can now be applied.*

Figure 4-6 shows how we want to assemble our copper fittings for this particular project. Use a *repair coupler* to get the tee installed in line. A regular coupler has stops in the middle of the fitting so that the pipe can only be inserted half way. A repair coupler has no stops, so the coupler can slide all the way past the end of one piece of pipe and then slid back down over the end of the pipe to be joined.

The key to sweating copper is to have all your pipe and fittings absolutely clean and dry. Use a wire brush, emery cloth and rags to get everything shined up.

Shut the water off at the meter and also shut off the valve at the foundation where the waterline enters the house. When you cut your waterline, all the water in the pipe is going to drain out into the hole you have dug. You at least need to get the water shut off at the meter. It will be helpful to locate the second shut-off where the waterline enters the house through the foundation or basement wall. If you can locate and close that valve then the house piping will not drain back out into the hole you have dug when the pipe is cut.

It's good to have enough space dug out below the pipe so that the water has someplace to go without submerging the cut waterline. This will keep dirt from sucking back into the line. Also, if there is any chance of the water in the house piping running back out the cut waterline, then you should shut off your electric water heater to avoid the possibility of burning out a heating element while the water is off.

Pipefitting skills are discussed in chapter one of this book in the section on pipe.

Once you have your tee and male adapter sweated onto the water line, Give the tee a quick flush by turning the meter on for a few seconds. Then, put 2 or 3 wraps of Teflon tape on the male threads and install your shut-off valve. For this project we are using a 3/4" brass angle valve, but you could certainly use a different valve for your own project if you had a preference. Gate valves and ball valves are also suitable for this application. Refer to

4-8. An angle valve installed on a copper waterline to serve as the irrigation main valve. Now that this valve is installed, the water meter can be turned back on to supply the house. This valve now isolates the sprinkler system from the rest of the water supply.

Chapter 1 for a review on valves.

You will now have some air in your waterline, so close the sprinkler valve you have just installed and open up one of the outside hose bibs on the house. With the hose bib open, open the water shut off at the house and then slowly open the water meter valve. Let the water run from the hose bib for a few minutes until the water runs clear and then shut the hose bib off. Check for leaks. ***Everything must be absolutely drip tight.*** Of course, it is, so take a break. (If it isn't, take a break anyway and then try again).

With your main shut-off valve installed and your house water turned back on, you can relax a little. When you want water for the sprinkler system, all you do is open this valve you have just installed and there will be no need to interrupt your water service again. It's a beautiful thing.

POINT-OF-CONNECTION / GALVANIZED PIPE

Some older homes still have galvanized steel water lines. To install your tee in this type of pipe you need to dig a large hole like we did for the copper pipe. Again, inspect the exposed pipe carefully before doing any cutting. Refer to Chapter 1 for more information on galvanized steel pipe.

COMPRESSION FITTINGS

There is an alternate method for getting a tee in either a copper or galvanized waterline, and that is by using a plastic compression tee. A *compression tee* is relatively easy to install and does not require the pipefitting skills of the other methods we have discussed. This fitting simply slides on the pipe where you have cut the line and then the ends are tightened with a wrench. The ends tighten up against a rubber washer that makes a drip tight fitting. Some installers scorn this method of installing a tee on the grounds that it could develop a leak as the rubber deteriorates after many years in the ground. Also, ground shifting and settling could stress the joints and cause a leak. I have asked plumbing inspectors about this type of connection and some will accept it, and some will not. My own preference is to not use compression fittings on a mainline. But, if you choose to make your point-of-connection this way, check your local plumbing requirements first to make sure it will fly. If you do install a compression fitting on a galvanized pipe, make sure to clean the outside of the pipe until it is smooth before attempting to install the tee.

BACKFLOW DEVICE

Now that we have our main valve installed, we can install our backflow preventor. On this particular project we are using a backflow preventor called a double check valve assembly or DCVA. Local code will apply to this installation. Ours will be installed with the top of the device 24" below grade, test cocks facing up, and schedule 80 plastic unions at both ends.

Check locally to determine permit and installation requirements that will apply to you. Some inspectors will want to see the installation from the point-of-connection to the backflow device, so you will not want to backfill that section of piping until after it is inspected. Pipe will have some lettering stamped on one side that indicates the grade of pipe. Install pipe that is to be inspected with the lettering facing up.

Make sure that you have the backflow device installed in the right direction. All backflow devices and many valves have a direction of flow. This direction of flow will be indicated with an arrow indicating the correct direction of the flow of water. Another way that the direction of flow might be indicated is by having the word "in" on the inlet side of the device and the word "out" on the discharge side of the device.

After the DCVA is installed, we're going to cover it with a plastic valve box that is large enough to provide room in the box to remove the check valves from the device for servicing. Our box will also have adequate clearance to get a pipe wrench on the unions so that we can completely remove the device without doing any digging. We will also provide about 6" of space under the DCVA, and line the valve pit with gravel for drainage.

4-9. A "compression tee" allows a tee to be installed in a copper or galvanized pipe without the pipefitting skills required by other methods. The compression tee slides on the pipe, and then the end caps are tightened, forming a seal against the rubber washers.

4-10. Angle valve connected to a compression tee with a schedule 80 nipple.

The DCVA is one of the backflow devices that needs to be tested when it is installed. You will need to call a certified backflow tester to do this for you.

DRAIN VALVE

If we are in a climate where freeze protection in the winter is a concern, we will install a manual drain valve to drain the mainline. For this project I'm going to use a 3/4" brass angle valve. As in most things we do, it doesn't have to be this particular type of valve. I chose this one because it will be sturdier than the 1/2" valves that are often used for drains, and because it will be available with a "cross" handle that can be turned from above with a long handled key. Because this valve will be located 2' below grade, we need to preserve access to it so that we can open and close it manually. We will cut a 2' length of 2" diameter PVC pipe to slide over the top of the valve before backfilling. Then, we will cover the top of the 2" pipe with a hinged brass cap made for this purpose. The cap will keep dirt and debris out of the PVC sleeve.

There are also automatic drain valves available. These valves consist of a diaphragm that is held closed by water pressure when a line is charged, and then opens when the line pressure drops below a certain point. We're not using any automatic drains on this project, because I don't see any application here where this type of drain would be appropriate.

Any drain, manual or automatic, needs a sump to drain into. A fence post digger is a good way to make a sump 12" deeper than the discharge of the drain valve. Fill the hole with drain rock.

QUICK COUPLER VALVE

Just after the drain valve we will also install a quick coupler valve. This is optional. I like it because it's a convenient place to hook up the air hose for winterizing the system with compressed air. When I do install a quick coupler, I like to put it on a double swing joint so that if it gets yanked by someone attaching a hose it will give at the joints instead of breaking off at the pipe. Quick couplers on swing joints can be "folded" into a valve box, or simply installed with the top of the valve above grade. See Figure 1-7 for a quick coupler assembled on double swing joints.

We will assemble our swing joints using 3/4" schedule 40 plastic street ells MIPT/FIPT. The designation "MIPT/FIPT" stands for "male iron pipe thread x female iron pipe thread" which means that one end of the fitting has a female thread and the other end has a male thread.

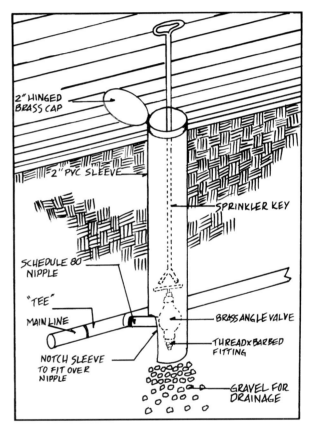

4-11. An angle valve installed "backwards" to serve as a manual drain valve. The valve is connected to a PVC mainline with a tee and a schedule 80 nipple. The valve is covered with a 2" diameter PVC sleeve to provide access from above, so that the valve may be operated with a sprinkler key. The valve remains closed during the operating season of the sprinkler system, and is only opened in the fall to winterize the sprinklers. When the valve is open, it drains into a gravel sump. The thread x barbed PE fitting, threaded into the discharge side of the valve where it enters the sump, is to prevent rocks from getting up in the valve. Note that this angle valve is installed to flow in a reverse of normal direction when opened. Use only a valve with a "rising stem" for this. A valve without a rising stem would leak if installed backwards. Check with your supplier for product suitability for a specific application.

When we assemble these street ells into swing joints we want to achieve a joint that is loose enough to swivel and tight enough to not leak. I've gotten good results by wrapping 2 turns of Teflon tape on the male part and applying pipe dope on the female part. This allows me to achieve a fit that can be relatively loose for good joint movement while remaining leak free. It is important to use a pipe dope that is compatible with PVC plastic.

There are pre-assembled swing joints available that seal with an "O" ring. These also work well for installing the quick coupling valve.

❧ Mainline and Wires ❧

WIRES

Our next step is to install the mainline and wires. Mainline and wires go together like peanut butter and jelly. That is because the mainline has to go to each valve and so do the wires. So, where you have one, you very often have the other.

Wiring is not that difficult, but it does seem to intimidate people. I think it will be helpful to you to understand some basic concepts.

The electric control valves that operate your sprinklers open and close by the use of an electric solenoid. The solenoid that comes as part of each electric control valve has two short wires attached to it. One of these wires is "hot" and the other wire is "common." These two wires are not marked because you may use either as the hot or the common.

This solenoid also contains a plunger that lifts when 24 volts of electrical current flow through it. This lifting of the plunger opens a small port, which bleeds water off of the top of a diaphragm, which in turn causes the valve to open and then the sprinklers go on. As long as there is electrical current flowing through the solenoid, the valve will stay open. When the electrical current stops, the plunger inside the solenoid returns to a position where it acts as a stopper covering the exhaust port, and the valve will return to a closed position.

The 24 volt AC current which activates the solenoids is provided by a transformer which turns your 117 volt AC household current into the low-voltage (24 volt) current required by the solenoids. This transformer is in integral part of the controller (or timer) that you install to run your sprinkler system. Besides converting high-voltage to low-voltage, the controller also has the ability to send the voltage to a specific solenoid for a specified time. This is what allows the valves to come on at a predetermined time and to stay on for as long or as brief a time as you specify when you program the controller.

Every controller will have some kind of terminal strip, which is a place to connect wires.

For a sprinkler system with six valves, you will buy a six station controller. A six station

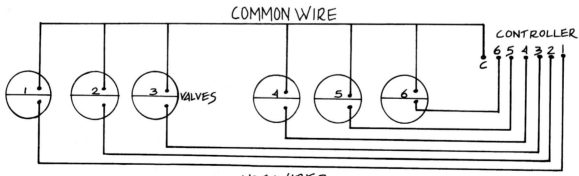

4-12. A schematic wiring diagram for a six station sprinkler system. One wire from each valve solenoid is joined to a common wire which connects to the common post at the controller. The other wire from each valve solenoid connects to its respective station post at the controller. The electrical current flows from the controller to the solenoid via the "hot" wire and returns via the "common" wire to complete an electrical circuit.

controller will have a minimum of seven places to connect wires. One post on the terminal strip will probably be labeled "C" or "COM". This is where you connect a common wire. The common wire is a wire that is connected to one wire of every solenoid in the system. ***The common wire is always white in color.*** This is a convention that is very helpful if you ever need to troubleshoot an electrical problem.

The other 6 posts on the terminal strip will be numbered 1 through 6. These posts are for the "hot" wires. Each solenoid has a hot wire that goes from the solenoid on the valve, to the respective post on the controller terminal strip.

Our project is using a six station controller because we need room to wire in six electric control valves. If we had twelve electric control valves, we would need to get a twelve station controller, and so on.

Current from the transformer goes out the hot wire to a particular solenoid, and returns via the common wire to complete an electrical circuit.

Other posts that may be found on some controllers provide for attaching a wire to go to a pump start relay, and to go to an electric master valve.

The pump start feature will only be used if you have a pump somewhere in your system.

An electric master valve is used if you do not want to have the mainline charged when the sprinklers are not running. With an electric master valve, the master valve opens simultaneously when any control valve is open and is closed the rest of the time. This feature reduces the possibility that you could ever lose water from a mainline break when the sprinklers were off.

For the purpose of our example here, there is no pump in the system and we are not using an electric master valve. Therefore we will need only to concern ourselves with one *hot* wire from each valve solenoid going back to the controller, and the *common* wire which will connect all the valve solenoids to each other and to the controller.

For our project refer back to Figure 2-13. Starting at the manifold of (3) valves in the front yard that are furthest from the clock we can see that we will need four wires (one hot wire for each valve plus one common wire) approximately 180' in length each, to go around the house to the side of the garage where the controller is to be located. Then, we will need three more hot wires from the manifold of (3) valves in the back yard of approximately 90' each. This second bank of valves will use the same common wire as the front valves so there is no need to run a second common wire.

We can accomplish this by using seven individual lengths of 14-1 UF direct burial low voltage wire, totaling 1,080', or we can run one 180' length of 18-7 multi-conductor cable which contains seven individual wires in one sheath. If you have a single wire run of over 1,000 feet in your own project, then you will need to use 14 gauge or larger wire. For short distances, like we have here, the 18 gauge multi-strand wire is fine. I would prefer to use the multi-strand wire, because that way I'm only laying one wire in the trench, instead doing it seven times over.

Whatever wire you use, the wire goes in the trench first so that it is underneath the pipe. For our project, let's pull out a few extra feet of 18-7 wire at the first valve pit and then spool it out in the trench to the backyard valve pit, being careful not to knock the dirt back into our trenches as we go. When we get to the second valve pit we will pull out a loop of wire about 3' long so that we have something to work with later when we splice in our solenoid wires. From there, let's continue on to the controller location, carefully laying the wire on the bottom of the trench as we go. Notice that the wire between the valve pits will be under the mainline and the wire from the backyard valves to the controller will be under a lateral pipe because the mainline ends at the second valve manifold.

When we get our wire to the side of the garage where the controller is to be mounted, pull out enough wire to go whatever distance is needed to get through the wall and to the controller. Having the wire a little too long is O.K. here, as we will cut the wires to fit when we install the controller.

For right now, we are interested in getting the right number of wires installed in the trench underneath the mainline, and having a loop of wire pulled up at each place where we will later install valves. We will leave the wire like that for now. Once the wire is in the bottom of the trench, we can lay in the rest of our piping. When we later install the controller, then we will also connect the wire to the valve solenoids and to the controller.

Mainline

With the wire in the trench, we can now install our mainline. This is the 1" diameter, class 200 PVC pipe, that will run from the discharge side of our DCVA to the electric control valves. Lay the pipe out alongside the trench and glue up the necessary lengths by using the belled ends of the pipe as couplers. After the glue has had a chance to set, place the pipe in the trench and connect it so that it is continuous from the backflow device to the last valve pit. When you are doing this, remember to put in a "tee" at the first valve pit so we can connect our valve manifold later without having to cut the mainline again.

Where the mainline goes around a corner you may want to cinch the wires to the pipe at each side of the corner, but do not pull the wire tight. Except at corners, I don't recommend cinching the wires to the mainline. That is because if you ever have a reason to cut the mainline later (such as during a repair) you will want to be able to cut through the pipe without nicking the wires. Cinching the wires tight against the mainline makes repair much more difficult, and doesn't really accomplish anything.

When I install "tees" in a mainline that will be used later for installing valves, I like to tape the open part of the tee with electrical tape until I am ready to use it. This keeps dirt and rocks from getting in the mainline while it is being assembled. Even though we will flush the mainline with water before installing the valves, I don't want to take a chance on getting a rock in my mainline that may not flush out later.

Control valves

To make our installation neater and easier, we can assemble our control valves on *manifolds* and then install two manifolds of three valves each. Figure 4-13 shows the assembly details of a typical manifold. Note that we want to install our valves a few inches higher than our mainline so that when the valve pit is finished off with a valve box, there will be some air space for drainage underneath each valve. We want to install our valves this way because whenever the bonnet of a valve is removed for servicing, some of the water trapped in the mainline will inevitably run out through the open valve body. If the valves are sitting down in the dirt, they will become submerged, allowing dirt and debris to be sucked into the valve being serviced. Therefore, we want some clearance under our valves for drainage.

Before connecting the valve manifolds to the mainline, turn on the main valve and flush any dirt or rocks out. Lift the ends of the mainline out of the valve pit when you do this

4-13. A two valve "manifold". The valves are "ninetied up" so that they are higher than the mainline. When the valves are covered with a valve box, the box will be notched to fit over the pipes. Inside the box, there will be air space under the valves. If the valves were not installed above the mainline, they (valves) would be sitting in the dirt after the valve box was installed.

to avoid filling the pit with water. When you flush the mainline, you may want to tape a temporary extension of pipe onto the mainline with electrical tape to direct the flush of water away from your work area. Flush the mainline thoroughly before you install the valves, because even a small amount of dirt can foul an electric valve and cause it to fail.

LATERALS

Our next step is to install the *lateral* pipes. A lateral, you will recall from chapter one, is any pipe on the discharge side of the control valve(s). Usually, I will start by laying out the lengths of pipe alongside the trenches. Then, I'll glue the lengths of pipe together by the belled ends. While the glue sets, I'll walk the project with the fittings and drop the appropriate fitting (tee, ell, and so forth) where it will be used. Then I'll go around with saw, glue can and rag and glue everything up. I like to start at the end of each line that is farthest from the control valve, and work back toward the valve pit. After all the laterals are assembled, I'll take the various pipes coming into each valve pit and plumb each lateral to the appropriate valve.

FLEXIBLE PIPE

There are different ways to attach sprinkler heads to the lateral pipe. They all involve some kind of transition fitting to connect the head to a tee. In my opinion, the best general purpose way to attach heads to any hard pipe is a flexible 1/2" PE pipe. This flexible pipe is manufactured under different names by different companies, but it all does the same thing. It is a 1/2" diameter, flexible soft pipe that uses fittings that are threaded on one end and barbed on the other to connect a head to a threaded tee.

When we assemble our lateral lines, it's easy to install the ells (for the PE pipe) in the tees before the tees are even glued into the line. Screw these plastic fittings into the tees until they are barely finger tight. They should not be so tight that you cannot move them. No Teflon tape or pipe dope should be used.

After we have finished gluing up the lateral lines, the next step is to push a length of flexible 1/2" PE pipe over the barbed end of the ells that have been installed in the tees. The flexible pipe is easily cut to length with hand pruners, a pipe cutter or your hacksaw. Cut a length that is longer than enough to reach from the tee to the location of its' respective sprinkler head, because after all the flexible pipe has been installed on the tees, we are going to backfill our trenches with the flexible pipe sticking out. Plan on having at least 6" of flexible pipe sticking up above grade after the trenches are backfilled. That way, we can give a good flush of water through the laterals before we put the heads on. When it is time to set the heads, we will trim the flexible pipe to length.

VALVE BOXES

The function of the valve boxes is to provide access to the valves for servicing. You don't ever want to install a valve in such a way as to make it not readily accessible. Do not bury the valves, pour concrete over the valves, build a deck over the valves or anything like that. Valves will require occasional service and always need to be accessible.

The valve box you select should be large enough to provide adequate clearance for someone to get their hands in the box with a small wrench to remove the top of the valve or unscrew the solenoid. A 10" round box is a good size for a single 1" valve. You can fit up to three 1" valves in a standard 12" rectangular box. When I see valves crammed into an undersized box, (or drain pipe, or coffee cans), I know I'm looking at a sprinkler system that wasn't installed professionally.

We don't want to set our valve boxes with the bottom edge of the box resting on top of the pipes for two reasons. One, having the edge of the box resting on a pipe could result in a broken pipe if anything heavy rolls over, or steps on the valve box. Two, dirt will infiltrate the box from the bottom and cover up the valves eventually.

Instead, we want to cut notches in the bottom edge of the valve box with a hacksaw,

4-14. A single valve installed in a valve box. The box is notched, and the valve is higher than the mainline.

so that the box straddles the pipes, with the weight of the box resting on the dirt and not the pipes.

If you are working in soft soil or fill that is going to settle, tamp the soil before setting the box so that it will not settle later.

We want the top of the box to be level with the finished grade. Extensions are available to extend the depth of the box if your valves are deep.

When you set your valve boxes, make sure that you have the wire for the valve solenoids pulled up into the box so that we can splice the solenoid wires into our field wires later.

Valve boxes will be much less obtrusive visually if you set them parallel with, or lined up with, something else on the site such as a sidewalk or wall.

You may set valve boxes in either lawn or shrub bed areas. When a valve box is in a lawn, it is important for the box to be very flush with the grade so that a mower can go over it.

Backfilling

When we're backfilling our trenches, we want soil with no rocks in it around the pipe. A trench backfilled with loose soil will settle later, so compact the soil by walking on it as you work. Soaking the trenches with water will also help settle the soil.

Set heads

By now, the bedlines you painted out at the beginning of the job will be scuffed off, and most of the flags will have been knocked down or moved. Go ahead and paint the bedlines again and re-flag the head locations.

Go to your first zone and turn on the control valve for a few seconds, or as long as it takes to flush all the dirt out of the lateral line and the PE pipe that will be sticking out of the ground. If the controller is not installed yet, as in this project, "bleed" the valve on manually. When you buy your valves, have the salesperson show you how the manual bleed works on that particular valve.

After the lines have been flushed, we can set the heads. Figure 1-12 shows a typical way PE pipe is used to connect a sprinkler head to a PVC lateral.

Some general guidelines for setting sprinkler heads:

1. Set heads 1.5" to 2" away from walks and driveways. If a head is to close to a hard surface it will get hit by the edger. If the head is set too far in from the edge there will be a dry area behind the head.

2. Whenever possible, if a lawn is adjacent to a shrub bed, locate all the heads in the shrub bed. The lawn heads can be in the shrub bed, 2" in from the edge, and spray out at the lawn. This will reduce the exposure to mower and edger damage, and eliminate dry areas behind the heads.

3. When watering close to a building, have the heads set 6" from the walls and spraying away from the building. When watering along a fence, have the heads spraying away from the fence whenever possible.

4. Keep heads at least 12" away from windows to keep drifting spray from spotting the glass.

5. Set your heads so that they will be flush with the finished grade. If you are going to add 2" of bark to a shrub bed, then the heads

should have 2" exposed above the soil after they are set. If the head is on a slope, set the head at an angle to match the slope.

6. Once the head is in position, pack the dirt around it tightly with your heel so that it will stay in position.

After you have set all the heads on the first zone, go on to the second zone and repeat the procedure until all the heads are set. Remember to *always flush the pipes* before you set heads.

Nozzles

It doesn't really matter if you install the nozzles on the heads before or after the heads are set. After all the heads are set and the nozzles are on, you will want to run each valve and make whatever adjustments are necessary to fine tune the nozzles. Most models of spray nozzle and rotor nozzle have an adjusting screw to control the distance of the spray. The direction of the spray can be adjusted by either turning the sprinkler body or the flow tube on a spray head, and most rotors have some way to adjust the degrees of arc in their rotation. After seeing your sprinklers work, you may even decide to move some heads or change some nozzles to achieve better coverage.

After all your hard work up to this point, take the time you need to fine tune and adjust the nozzles now. It can make all the difference in terms of achieving good coverage and minimizing overspray.

Controller

Now we will go into the garage and install the controller. An indoor controller will typically have an external transformer that simply plugs into any convenient 117 VAC standard household electrical receptacle. We are usually going to want to mount the controller on a wall within a few feet of the electrical receptacle, so that we can plug the controller in. An outdoor controller will have an internal transformer, and will be enclosed in a weatherproof cabinet. To install an outdoor controller, you may need to provide a "J box" and splice the transformer wires into your power source.

The electrical current from the transformer to the valve solenoids is low-voltage. On the input side of the transformer, however, is high voltage current. Treat the electrical part of the project with respect, and get professional help if you need it, especially if you are going to have a hard-wired connection to the power source.

Mounting the controller on a finished wall usually requires only a few drywall screws. Mounting instructions should be furnished with your controller when you purchase it. If you plan to install the controller on a masonry surface then, of course, you will need the appropriate fasteners.

For our project, we will mount the controller inside the garage. We won't plug it in until all the wires have been connected at the controller and at the valves.

Connecting the Wires

After the clock is secured on the wall at a convenient height, we can take our power drill and drill through the wall. The field wires are all low-voltage (24 VAC) and we don't need to use any conduit unless specified otherwise by code. Since we used multi-strand 18-7 wire, we only have one small cable to work with about the size of a phone line. If I were using individual 14 gauge wires instead of multi-strand, then I would want to hide the wires in a 1" conduit for the sake of having a neat installation.

For our project, all we need to do is drill a small hole just large enough to push our cable through. We'll do this as low as we can, just above the floor joist. Now we can push our wire through and then run it up the inside wall and into our controller. Strip the wire ends and attach the wires to the terminal strip. The white wire is always the common wire, so attach that to the terminal marked "com". You don't really need to be concerned about a color code for the rest of the wires as long as you have the wire from the valve you intend to be number one connected to station one at the controller, and so forth.

Now go back to your valve boxes and

wire the valve solenoids. At the valve box closest to the controller you are going to carefully slice the outer insulation of our cable and peel it back. Pull out the seven small insulated wires and cut the white one and three other colors. Strip about 1/2" of insulation from the ends of these four wires. Strip the end of the wire that leads back to the controller only. In addition, also strip the end of the white wire that leads on to the next manifold of valves. Each solenoid in the valve box has two wires. Select one wire from each solenoid and secure it to one of the colored wires with a wire nut. Now take the remaining wire on each of the three solenoids and join those with both ends of the white wire. If the solenoid wires are not long enough to connect all the commons together, you can use an additional length of wire as a "jumper". Take the remaining ends of the three colored wires that were cut, but not spliced to anything, and wrap them around the outside of the cable.

Now, go to the remaining valve box in the front yard and peel back the end of the outer insulation on the wire there. There should be a small string that you can pull to cut through the outer insulation, exposing the seven inside wires. Pick the three remaining unused colors and attach one of these to one wire from each solenoid. Then, attach the one remaining wire from each solenoid to the white wire. Wrap the unused wires around the outside of the cable. You should now have one "hot" wire from each solenoid spliced to a field wire that is connected to a station terminal post at the controller. The other, "common" wire from each solenoid should be spliced into the white wire that is connected to the common terminal post at the controller.

Waterproof splices

Wire nuts will not protect your splices from moisture and neither will electrical tape. There are various waterproof splice containers available and you should insert your wire nuts into one of these products to seal out moisture. Check with your supplier to see what is available. I prefer the plastic containers that only require you to insert the wire nut, push the wire in an inch or so, and snap the end shut. Quick, clean and easy.

After you have all the wires wire nutted together, but before you put on the waterproof containers, check the operation of the controller and valves. If you want to switch some wires around, it will be easier if you don't have to pull the waterproof containers off. Put the wire nuts in the waterproof containers after the controller and valve operation have been checked and you know you have the system up and running.

The test drive

The next step is to plug in the controller and program it according to the manufactures directions. I know you will be anxious to try out the system, but take your time here and read the directions carefully. Follow the

4-15. A four valve manifold. The solenoid wires are connected to 18 gauge multi-strand wires. The splices are done with wire nuts that are then enclosed in waterproof containers.

directions to the letter and don't skip any steps. Once the controller is programmed, turn on station one. This is the part I like. Again, I'm not a gambler, but if I was I'd be willing to bet you'll be wearing a rather large smile when you push the button on the controller and station one comes on for the first time.

TROUBLESHOOTING

If your sprinkler system does not perform flawlessly the first time you start it up, don't worry.

A new sprinkler system almost always needs some tweaking to run right, or to run at all for that matter. Do you remember when you brought home your first computer and took it out of the box and plugged in all the peripherals and tried to get everything to work? If this is your first sprinkler system, the odds are that you will make some adjustments to get it up and running correctly. Here are some common maladies and their solutions:

1. You push the manual start button on the controller and nothing happens for any station.

A. Check to make sure that the mainline is on. Bleed the valves on manually. If you can bleed the valves on manually, then you know it is an electrical problem and not a water problem.

B. If you have determined that it is an electrical problem, first check that the controller is programmed correctly, that you have time on the stations, and the fuse is not blown. Also, make sure it's plugged in.

C. Check the common wire. Since this is probably your first sprinkler system, make sure that you have the valves wired correctly. A bad hot wire will only cause one station to fail. If all of the valves do not work from the controller, the common wire is not installed correctly or is broken. Check your splices.

D. Check the direction of flow at your control valves. You didn't install the valves backwards, did you?

E. If you could not bleed on the control valves manually, then the mainline is not charged. Make sure that the main valve is open, and also that the ball valves on the backflow device are open.

2. One valve does not work from the controller.

A. Check at the valve to make sure the flow control is not turned down.

B. Check the wire connections at the valve and at the controller.

C. Check the programming on the controller. Make sure the station in question has time on it, and is on the program you are using.

3. The valve works, but one or more heads on the zone do not pop-up.

A. Dirt or rocks plugging the connection to the head. Dig back to the offending connection and clean it out, or cut it out and replace it.

B. Kinked PE pipe.

C. Broken pipe or fitting. The water should surface if this is the problem. Dig it up and fix it.

D. Is it possible that the head that appears to be not working operates on a different zone?

4. The head pops-up, but does not spray.

A. Dirt in the nozzle. Remove and clean the nozzle and screen. With the nozzle removed, blow some water through the head to clean it out before replacing the clean nozzle and screen.

B. Adjusting screw in the nozzle turned down. Back it off with a small screwdriver.

5. The heads spray, but weakly.

A. Check that manual and electric valves are not flowed down.

B. Check for broken pipe causing loss of water. Repair.

C. Check design. Are you trying to run too many heads on one valve? Is the pipe size too small? Refer to the design section in this book.

D. Obstruction in pipe or valve. If this is the case, you can deduce that the obstruction will have to be ahead (upstream) of the first weak head.

Be methodical. Calm down and think it through. If you have followed the directions up until now, your troubleshooting will likely be something simple that you have just overlooked.

When you get your sprinklers working, check out the next chapter which will explain how to get the most out of your sprinkler system by understanding how to schedule the watering cycles in a way that is appropriate to your particular site.

By the way, congratulations on your accomplishment!

Chapter 5

Using Your Sprinkler System

In Using Your Sprinkler System *you will learn something about the relationship between plants, soil and water. You will find out how these elements are factors in a watering schedule. You will also review some of the features found on automatic controllers and how to use those features to create an appropriate watering schedule for your own landscape.*

This will enhance the value of owning a sprinkler system. By understanding how to water, you can get the most value possible from your sprinkler system, save water and money, and make your green thumb even greener.

SOIL

Some understanding of your yard's soil will be helpful to you because your soil condition determines the percolation rate of the water you apply and affects the root depth (and water storage capacity) of the plants you grow.

The percolation rate is the rate at which water applied to the soil can be absorbed into the soil. The reason why this is significant is because if the sprinklers apply water at a precipitation rate that is in excess of the percolation rate, then we will have puddling or run-off. We want to apply water at the same rate as the soil's capacity to take in the water.

The three very broad categories of soil are *clay, loam* and *sand.*

Clay is soil that is composed of very small particles. If you pick up a handful of clay and form it into a ball it will hold its shape and not crumble when you open your hand. Clay is slick and shiny when wet. It is often referred to as a "tight" soil because water does not percolate into it well.

When you are watering plants in clay soil, you want to use sprinklers that apply water at a low precipitation rate. It is easy to apply water to clay soils at a rate that is faster than the soil can absorb, causing excessive run-off.

In addition choosing low precipitation rates in your sprinkler design, you can adjust the watering time on your controller to help conserve water by keeping your run times short. If you use a controller with multiple start times, you can use that feature to water several times a day for short periods instead of one longer period. By using several shorter run times you only apply as much water as can be applied until run-off occurs, and then give the water time to slowly percolate into the soil before watering again.

For example, when watering a lawn in clay soil that is irrigated with spray heads, two start times of five minutes each may keep the lawn just as green as one 20 minute soaking and will use half as much water.

Sand is soil with very large particles. Unlike clay, sand cannot be formed into a ball with your hand, because it crumbles as soon as your hand is opened.

When water is applied to sandy soil it does not run-off, but instead percolates rapidly down through the course particles.

In order to be effective, water applied to plants needs to be in the soil at the depth where the plant roots are. Sandy soil will suck up large amounts of water, but that water will be wasted if it all ends up deep in the soil below the roots.

To compensate for sandy soil, like clay soil, use multiple start times. In clay soil we want to stop applying water before it runs-off. In sandy soil we want to stop applying water before it reaches the soil that is deeper than the plant roots. These are extremely different soil conditions, but the watering solutions are very similar- low precipitation rates, short run times, and multiple start times.

Loam is a soil that is a medium texture. If you pick loam up with your hand and form it into a ball, it will maintain some shape when your hand is opened and will crumble when touched. A loamy soil with a lot of composted organic material tilled into it is the best planting medium and the easiest to water. Many watering problems that occur are actually soil problems.

I don't think it is any exaggeration at all to say that good soil preparation for your planting is a key ingredient to installing a successful sprinkler system. No amount of hardware and plumbing expertise can compensate for poor soil.

A good soil will provide space for air, water and roots. A loose, deep soil will encourage deeper root growth and increase the ability of the plants to store water. When the roots are deeper, you can water deeper.

WATERING

Most plants do not like extremes of wet or dry.

When you saturate the soil, you fill up all the space between soil particles with water and the plant roots can't exchange nutrients and oxygen with the soil. When you let the soil dry out, the fine root hairs dry and shrivel and the plant looses its ability to function. For this reason, you should avoid watering in a cycle of soaking and drying. It is better to create a slightly moist soil at the root depth and then water again when the plant has used approximately 50% of its stored water.

In other words, stop watering before the soil is completely saturated (run-off occurs) and water again before the soil is completely dry at the root depth. To figure this out for your particular site, you will have to spend some time experimenting with a soil probe to see how deep the roots are and how your soil moisture is at different depths after different run times with your sprinklers. Once you have done this, you will have a much better feel for how long and how often to run your sprinklers.

Moisture you apply to your plants is lost by a process called evapo-transpiration or ET. Evapo-transpiration is a combination of evaporation from the soil and transpiration through the leaves of the plants. The ideal amount of water to apply is equal to the amount of water lost to ET every day. The ET rate is created by a combination of solar radiation, wind, temperature and relative humidity.

PLANT COMMUNITIES

Different plants have different watering needs. When we designed our sprinkler system, we zoned it so that plants with similar water requirements were served by the same valve. By doing this, we can apply the right amount of water to all the plants. In our landscape design, we can avoid mixing plants in a grouping if the plants have very different water needs. This will eliminate many vexing irrigation problems.

TIME OF DAY

The best time to water is early in the morning. This will minimize the water that is lost to evaporation. Watering in the evening is generally not advisable because water left on the foliage all night encourages some kinds of fungus to grow.

Schedule your irrigation to run before or after your households peak morning water use. For example, if the total run time for all the zones on your sprinkler system is one hour and your family gets up at 6 A.M., then have the sprinklers come on at 5 A.M. and you will have all of your water available for household use when you need it.

There may be times when you want to cycle the sprinklers later in the day to keep the soil moist, such as after a lawn renovation or new lawn installation, a new planting, or a fertilization during hot weather. You will find your automatic sprinklers to be extremely helpful in these situations if you purchased a controller with a multiple start time option.

WATERING DAYS

What days to water depends on the type of plants and the storage capacity of the plant roots. Generally, a new planting will need to be watered more frequently than an established planting. If you landscape with native or naturalized shrubs and trees, then you will find that very little supplemental water is required once the plants are established. A new lawn, on the other hand, will need to be watered every day for a few weeks until it is established.

Some communities have watering restrictions that will determine what days you may water. In that case, you will find that an automatic sprinkler controller helps to keep a watering schedule for you, and you won't have to plan your life around your watering days in order to be home to drag the hose around on your assigned days.

MULTIPLE PROGRAMS

Many controllers have a multiple program feature. This is very useful for being able to tailor a watering schedule to different plant groups. For example, if I want my lawn to be watered four times a week, but I only want my shrubs to be watered twice a week, then I will use one program for the zones that water the lawn and use another program for the zones that water the shrubs. If a controller does not have a multiple program option, then all of your zones will have to water on the same days.

Water time

All controllers have some way to set the amount of time each station waters. This feature lets you pick the number of minutes that is appropriate to water a group of plants each time that particular valve comes on. For example, if I have a lawn being watered by spray heads, I might set that station to water for 10 minutes. The next station might be a zone of established shrubs watered by spray heads, and I might have that station water for only 4 minutes.

Water budget

Some controllers have a water budget feature that allows you to change the station water times across the board without re-setting each individual station. For example, I could set my water budget for 120% and the time on each station would automatically increase by 20%. Or, I could set my water budget to 90% and the time on each station would automatically decrease by 10%. This is a handy feature. Once you get your base run times established, and the run times for each zone are correct relative to each other, then the water budget feature is a convenient way to compensate for fluctuations in the weather. when you get a cloudy week you can trim the water back, and when the temperature rises and the humidity drops, you can bump the time up a little by only changing one setting. This is much easier than having to change every station every time the weather changes. In fact, without this feature most people won't change the settings on their controller as often as they should.

Moisture sensor

A moisture sensor, correctly installed, will interrupt the watering schedule whenever adequate moisture is already available from rainfall. This feature eliminates the need to manually switch the controller to a standby mode when it is raining. There are also sensors available to interrupt the watering schedule when the temperature drops below freezing.

The future

In the very near future you will have the option of having your sprinkler controller linked to the computer at a landscaping company. Many commercial and public sprinkler system owners already are on systems like this. The landscaping company is linked to a weather station and your sprinklers are adjusted every day, based on current weather conditions. Your sprinkler system can also be fitted with flow sensors so that the watering is automatically interrupted if a sprinkler head or pipe is broken.

The benefit of a system like this is that attractive landscapes can be maintained with much less water than was ever possible before. The cost of these systems has been prohibitive for homeowners in the past, but as the price of the technology comes down and the cost of water goes up, my opinion is that we're about to see home irrigation technology take a big jump forward. The links from a central computer to the controller in your garage can be via phone line, radio, or cellular phone, so the infrastructure is already in place, and the technology is already being used very successfully by large water users.

When will you be using it? That's hard to say, but for now just having a sprinkler system is a big step up from hand watering. That's not to say that hand watering can't be a pleasant way to spend a summer morning in the garden. It's just that, day in and day out, most of us don't have the time or the inclination to really do a good job of handwatering. And there is no way hand watering can ever approach the efficient water use that a sprinkler system can achieve.

A well designed sprinkler system will do a superb job of watering, and will require very little attention once you get past the initial learning curve.

Enjoy!

Appendix

Pressure Loss Tables and Velocity of Flow Tables

The tables provided in this section will help you to design your sprinkler system. You can use the tables to find the numbers you will need to do a "worst head" analysis, and to size your pipe and valves to the correct flow and velocity. The numbers used here are approximate because the exact pressure loss through a component will vary slightly from one manufacture's product to the next. The most accurate information will be found in the manufacture's specifications for the specific products you are using. All flows are given in gallons-per-minute (GPM). All velocities are in feet-per-second (FPS). All pressures are in pounds-per-square inch (PSI).

CLASS 200 PVC PIPE. PRESSURE LOSS FROM FRICTION PER 100 FEET OF PIPE.

Flow (in GPM)	1/2"	3/4"	1"	1-1/4"
1	.26	.07		
2	.89	.26		
3	1.86	.52		
4	3.12	.90	.28	
5	4.76	1.37	.42	
6	6.62	1.90	.59	
7	8.82	2.52	.80	
8	11.26	3.21	1.02	.31
9		4.05	1.24	.40
10		4.90	1.52	.50
11		5.82	1.81	.59
12		6.83	2.14	.69
13			2.51	.80
14			2.86	.90
15			3.26	1.02
20			5.53	1.76
25				2.67

TYPE L COPPER TUBE. PRESSURE LOSS FROM FRICTION PER 100 FEET OF PIPE.

Flow (in GPM)	1/2"	3/4"	1"	1-1/4"
1	.97			
2	3.51	.60		
3	7.42	1.26		
4	12.68	2.15		
5		3.25	.89	
6		4.55	1.25	
7		6.04	1.65	
8		7.72	2.13	
9		9.61	2.62	.86
10		11.66	3.19	1.05
11			3.81	1.25
12			4.47	1.46
13			5.19	1.71
14			5.95	1.95
15			6.75	2.21
20			11.50	3.76
25				5.69

POLYETHYLENE (PE) PIPE. PRESSURE LOSS FROM FRICTION PER 100 FEET OF PIPE.

Flow (in GPM)	1/2"	3/4"	1"	1-1/4"
1	.51			
2	1.79	.48		
3	3.77	1.00		
4	6.43	1.64	.46	
5		2.46	.76	
6		3.45	1.06	
7		4.59	1.37	
8		5.85	1.80	.46
9		7.29	2.23	.59
10			2.76	.70
11			3.27	.85
12			3.85	1.01
13			4.45	1.20
14			5.10	1.37
15			5.76	1.56
20				2.56
25				3.87

AGED STEEL PIPE. PRESSURE LOSS FROM FRICTION PER 100 FEET OF PIPE.

Flow (in GPM)	1/2"	3/4"	1"	1-1/4"
1	.91			
2	3.28	.83		
3	6.95	1.77		
4	11.81	3.01	.85	
5		4.55	1.40	
6		6.37	1.97	
7		8.47	2.62	
8		10.83	3.35	.88
9			4.16	1.10
10			5.06	1.33
11			6.05	1.59
12			7.09	1.87
13			8.22	2.17
14			9.43	2.48
15			10.70	2.82
20				4.80
25				7.26

PRESSURE LOSS THROUGH ELECTRIC CONTROL VALVES

Flow (in GPM)	3/4"	1"
3	2.5	0.5
5	2.9	1.0
10	3.8	2.6
20	5.1	3.6

PRESSURE LOSS THROUGH COMBINATION CONTROL/ANTI-SIPHON VALVES

Flow (in GPM)	3/4"	1"
3	3.6	3.1
5	4.2	4.0
10	5.7	5.4
20	8.6	7.3

PRESSURE LOSS THROUGH DOUBLE CHECK VALVE ASSEMBLY (DCVA)

Flow (in GPM)	3/4"	1"
12.5	4	
20		4

PRESSURE LOSS THROUGH PRESSURE VACUUM BREAKER (PVB)

Flow (in GPM)	3/4"	1"
13	4	
20		4

PRESSURE LOSS THROUGH ATMOSPHERIC VACUUM BREAKER (AVB)

Flow (in GPM)	3/4"	1"
12	2	
20		2

PRESSURE LOSS THROUGH GLOBE ANGLE VALVE

Flow (in GPM)	3/4"	1"	1-1/4"
5	.54	.25	
6	.73	.32	
7	.94	.40	
8	1.22	.50	
9	1.52	.61	
10	1.88	.72	
11	2.29	.84	
12	2.74	.97	
13	3.13	1.15	
14	3.54	1.37	
15	4.04	1.64	.56
20		2.74	.94
25		4.24	1.44

PRESSURE LOSS THROUGH GLOBE STRAIGHT VALVE

Flow (in GPM)	3/4"	1"	1-1/4"
5	1.04	.47	
6	1.53	.59	
7	2.02	.78	
8	2.54	1.00	
9	3.07	1.24	
10	3.76	1.53	.52
11	4.54	1.82	.62
12	5.33	2.13	.73
13	6.21	2.44	.83
14	7.10	2.75	.94
15	8.13	3.24	1.10
20		5.40	1.90
25		8.10	2.84

PRESSURE LOSS THROUGH WATER METERS

Flow (in GPM)	5/8"	3/4"	1"
5	.9		
6	1.3		
7	1.8	.8	
8	2.3	1.0	
9	3.0	1.3	
10	3.7	1.6	
11	4.4	1.9	.8
12	5.1	2.2	.9
13	6.1	2.6	1.0
14	7.2	3.1	1.1
15	8.3	3.6	1.2
20	15.0	6.5	2.2
25		10.3	3.7

CLASS 200 PVC PIPE. VELOCITY OF FLOW IN FEET-PER-SECOND.

Flow (in GPM)	1/2"	3/4"	1"	1-1/4"
5	3.98	2.36		
10	7.97	4.72	2.89	
15		7.08	4.34	2.73
20			5.78	3.64
25				4.55

POLYETHYLENE (PE) PIPE. VELOCITY OF FLOW IN FEET-PER-SECOND.

Flow (in GPM)	1/2"	3/4"	1"	1-1/4"
5	5.28	3.10	1.86	
10		6.02	3.71	2.15
15			5.57	3.22
20			7.42	4.29
25				5.36

TYPE L COPPER TUBE. VELOCITY OF FLOW IN FEET-PER-SECOND.

Flow (in GPM)	1/2"	3/4"	1"	1-1/4"
5	6.88	3.31		
10		6.63	3.89	
15			5.83	3.83
20			7.78	5.11
25				6.38

STEEL PIPE. VELOCITY OF FLOW IN FEET-PER-SECOND.

Flow (in GPM)	1/2"	3/4"	1"	1-1/4"
5	5.28	3.01		
10		6.02	3.71	
15			5.57	3.22
20			7.42	4.29
25				5.36

Index

A

ABS plastic pipe 65
Air gap 18
Angle valve 7, 73
Anti-siphon valve 7, 9
Aquifers 31
Arc 4
Atmospheric vacuum breaker (AVB) 7, 9, 13, 18, 19, 21, 44
Automatic drain valve 7, 75

B

Backfill 66, 68
Backfilling 80
Backflow 7, 10, 11, 17, 21, 25, 31, 44, 57, 83
Backflow device 65, 78
Backflow preventor 74
Backpressure 19
Backsiphonage 19
Ball valve 7, 19, 20, 73
Barbed connector 56, 57
Barbed fittings 57
Base plan 32, 34
Basement connection 30
Bill of materials 65
Bleed screw 8
Bonnet 7, 78
Boring 69
Boring attachments 69

C

Catch cans 36
Caulking 67
Central computer 89
Certified backflow tester 75
Check valve 18
Circuits 43
Clay 87
Clay soil 32, 59
Common wire 77, 82, 83
Compressed air 7, 19
Compression fittings 15, 57, 73
Compression tee 74
Computer software 34
Container plants 62
Control valve 6, 18, 43, 68, 83
Control/anti-siphon valve 18, 21
Controller 8, 16, 46, 66, 77, 78, 81, 82, 83, 87, 89
Copper fittings 73
Copper pipe 12, 69
Critical path 66
Cross handle 6
Cutting die 14

D

DCVA 19, 21, 44, 74, 78
Design 25
Design Worksheet 51
Diaphragm 8, 76
Direction of flow 74
Discharge rate 34, 60
Disk holder 6
Distribution tube 56, 60, 62
Distribution uniformity (DU) 36
Double check valve assembly (DCVA) 19, 44
Double swing joint 75
Drain 75
Drain valve 6, 7, 75
Drainage 31, 78
Drainline 65
Drip 21
Drip emitters 58
Drip irrigation 34, 55
Drip line 59, 62
Drip tube 56, 60, 62
Drip valve 60
Dynamic pressure 28

E

Ear plugs 67
Electric control valves 44, 76, 77, 78
Electric master valve 77
Electric valves 8, 9, 56
Electro-mechanical controllers 16
Elevation 31, 48
Ell 14, 66, 79
Emery cloth 67
Emitter spacing 59
Emitters 56, 57, 60, 62
End strip nozzles 40, 42
Evapo-transpiration or ET 88
Exhaust port 8, 76
Extensions 80
External transformer 81

F

Feet per second (FPS) 46
Fertilizer injection 20
Filter 55, 60
Filtration 31
Fittings 57
Fittings 12, 14, 46, 66, 79
Flexible pipe 79
Flow 31
Flow rate 29
Flow tube 3
Flow-control 8, 83
Formula for calculating GPM 29
Freeze protection 29
Friction 29, 48
Fuse 83

G

Gallons-per-hour (GPH) 29, 57, 59
Gallons-per-minute (GPM) 29
Galvanized steel 13
Galvanized steel water lines 73
Garden valve 7
Gate and waste valve 6
Gate valve 6, 73
Globe angle valve 6
Globe straight valve 7
Gloves 67
Glue 11, 66, 79
GPM 36, 40, 43, 44, 46, 48, 50
Graph paper 32

H

Hacksaw 10, 12, 67
Hanging baskets 62
Hardpipe 56
Head layout 25, 39
Head selection 38
Head spacing 37
Head to head coverage 34
Head to head spacing 37
Heads 66
High voltage current 81
High water pressure 38
High-velocity nozzle 69
Hose bib 7, 28
Hose bib connection 57
Hot wire 77, 82, 83
Hybrid controller 16
Hydrozones 43, 44

I

Impact sprinklers 4, 5
Internal transformer 81

J

Jumper 82

L

Landscape plan 25, 32, 34, 39
Latching solenoid 8
Lateral piping 6, 79
Layout 67
Lead-free solder 67
leather footwear 67
Loam 87
Loamy soil 59
Low-volume (drip) irrigation 38, 55
Low-Volume Design 58

M

Main shut-off valve 5, 6,12, 73, 78, 83
Mainline 5, 6, 18, 21, 30, 76, 77, 78, 83
Male adapter 15
Manifolds 78
Manual valves 6, 55
Marking paint 34, 67
Micro-spray 55, 57, 58, 62
Moisture sensors 17, 89
Mounting the controller 81
Multi-conductor cable 77
Multiple port emitters 57
Multiple programs 88

N

Nipples 5, 10, 42
Nozzles 3, 4, 66, 81, 83

O

Overspray 38, 40, 42
Overthrow 38

P

Packing nut 9
PE pipe 11, 15, 79, 80, 83
Percolation rate 87

Pick 67
Pipe 10
Pipe cutter 10
Pipe dope 14, 66, 76
Pipe sizing 25, 46
Pipe threader 13
Pipe wrench 67
Pipefitting 14, 73
Plant communities 25, 88
Plumbing code 10, 20, 44
Plumbing inspection 66
Plumbing permit 20, 65
Point-of-connection 12, 25, 28, 44, 48, 68, 69, 73, 74
Polyethylene pipe (PE) 11, 56
Polyvinyl Chloride 10
Pop-up Sprinklers 3
Poppet 18
Power drill 67
Precipitation rate 32, 36, 43, 87
Pressure 29, 31
Pressure compensating emitter 57, 60
Pressure compensating filter screens 42
Pressure gauge 28
Pressure loss 48
Pressure reducer 50, 55, 60
Pressure regulator 21
Pressure vacuum breaker (PVB) 13, 18, 19, 21, 44
Primer 11, 66
Programming 83
Propane torch 67
Public water supply 20
Pump start relay 77
PVC 10, 11, 56, 57, 75, 76, 78, 80

Q

Quick-coupling key 7
Quick-coupling valve 7, 75

R

Rags 67
Ratchet 14
Reduced pressure principle (RP) device 20, 22, 44
Repair coupler 73
Resilient seat 6
Riser 3, 5
Root depth 87, 88
Rotor heads 42
Rotor nozzles 5
Rotors 4, 40, 43
Run time 60
Run-off 25, 32, 87

S

Safe flow 28
Safety glasses 67
Sand 87
Sandy loam 32
Sandy soil 59
Schedule 40 PVC 10
Schedule 80 PVC 10
Screwdrivers 67
Setting sprinkler heads 80
Shovel 67
Shrub adapter 5
Shrub heads 37
Shut-off valve 30
Side strip 40
Site analysis 25, 26
Sledge hammer 67, 69
Sleeve 9, 15
Slip joint pliers 67
Soaker hoses 57
Softpipe 56
Soil 32, 58, 87
Soil probe 88
Solder 13
Soldering flux 67
Solenoid 8, 76
Solid state controllers 16
Spaghetti tubing 56
Spray head nozzle 5
Spray heads 3, 42, 43, 58, 87
Sprinkler flags 67
Sprinkler head 3
Sprinkler head selection and layout 34
Sprinkler key 6
Stake 62
Static water pressure 26, 28, 29, 48
Steel pipe 14
Straight valve 7
Stream bubblers 42
Sweating copper tube 13

T

Tape measure 67
Tee 14, 60, 66, 78, 79
Teflon tape 66, 73, 76
Terminal strip 76
Test cocks 19
Threaded couplers 5
Threaded fittings 14
Tracing paper 32, 34, 39
Transformer 16, 76
Trencher 66

Trenching 68
Troubleshooting 83
Tubing cutter 12, 67

U

Underground utilities 65
Unions 15
Utility easement 26
Utility locates 65

V

Valve bonnet 8
Valve boxes 9, 44, 66, 74, 78, 79, 81
Valve manifolds 44, 78
Valve pits 9, 66, 68, 77, 78
Valve solenoids 82
Valves 5, 66
Vegetable garden 62
Velocity 29, 48
Vent 18

W

Water auditor 37
Water budget 89
Water hammer 46
Water meter 26, 32, 48, 73
Water storage capacity 87
Water time 89
Watering 87
Watering days 88
Watering restrictions 88
Waterline 30
Waterline replacement 14
Waterproof splices 9, 66, 82
Weather station 89
Wells 20, 30
Wheel handle 6
Wire 17, 66, 76, 78, 80, 81
Wire cutters 67
Working water pressure 28, 29
Worst head 50, 52
Worst head analysis 48

Z

Zone of reduced pressure 20
Zoning 25, 42

❧ ORDER FORM ❧

To order additional copies of *HOW TO DESIGN AND BUILD A SPRINKLER SYSTEM* complete and return this order form. Photocopy of this form is O.K.

Number of copies   _____

Times price each @ $ 19.95

Subtotal   _____

Add $ 3.00 shipping $ 3.00

TOTAL $_____   <u>Enclose check</u> with full payment payable to:
                                       "Irrigation Publishing"
                                       Do not send cash.

(Please print) My payment is enclosed. Please ship my book(s) to the following address:

Name_____
Street_____
City_____State_____Zip_____

(optional) Phone_____
E-mail_____

Return this form with payment to:

IRRIGATION PUBLISHING
Dept. B-1
P.O. Box 22184
Milwaukie, Oregon 97269-2184

(Allow 4-6 weeks for delivery)